Practice Test
for the
CogAT® Grade 5
LEVEL 11

ISBN: 978-1-948255-92-9

The Cognitive Abilities Test (CogAT®) is a registered trademark of Houghton Mifflin Harcourt, which is not affiliated with Origins Publications. Houghton Mifflin Harcourt has not endorsed the contents of this book.

Origins Publications
New York, NY, USA

Email:info@originspublications.

Origins Publications

Origins Publications helps students develop their higher-order thinking skills while also improving their chances of admission into gifted and accelerated-learner programs.

Our goal is to unleash and nurture the genius in every student. We do this by offering educational and test prep materials that are fun, challenging and provide a sense of accomplishment.

Please contact us with any questions.

info@originspublications.com

Contents

Part 1. Introduction to the CogAT® 4

Who Takes The CogAT® Level 11? 4

When Does the CogAT® Take Place? 4

CogAT® Level 11 Overview 4

Length 4

Format 5

Test Sections 6

Part 2. Using This Book 7

Part 3: Test-Taking Strategies 7

Sample Questions and Teaching Tips 9

Part 4: CogAT® Level 11 Practice Test 19

Verbal Battery 21
Nonverbal Battery 41
Quantitative Battery 69

Answer Keys 86
Bubble Sheets 94

Introduction to the CogAT®

This book offers an overview of the types of questions on the CogAT® Level 11, test-taking strategies to improve performance, sample questions, and a full-length practice CogAT that students can use to assess their knowledge and practice their test-taking skills.

Who Takes the CogAT® Level 11?

There are ten CogAT levels, which are based on age. The CogAT Level 11 (Form 7) was made for 11 year olds, and is often used as an assessment tool or admissions test in 5th grade for entry into 5th/6th grade of gifted and talented (GATE) programs and highly-competitive schools.

The CogAT Level 11 is also used as an assessment tool by teachers to figure out which students in grade 5 would benefit from an accelerated or remedial curriculum.

When Does the CogAT® Take Place?

This depends on the school district you reside in or want to attend. Check with the relevant school/district to learn more about test dates and the application/ registration process.

CogAT® Level 11 Overview

The CogAT is a group-administered test that features three independent 'batteries': Verbal, Quantitative, and Nonverbal. It is designed to assess learned reasoning in these three areas, which experts believe are the areas most closely linked to academic achievement. One, two, or all three batteries may be administered based on the specific needs of the test user.

The CogAT covers topics that students may not see in school, so kids will need to think a little differently in order to do well. A student's stress management and time management skills are also tested during the exam.

Length

Students are given about 30 minutes to complete ONE battery in the CogAT Level 11 test. Children can be given each battery separately or all at the same time. A test administrator provides directions and controls pacing throughout the test. Including administration time, the CogAT 11 (all three batteries) will take between approximately 2.5-3 hours.

Format

The entire test is made up of 176 multiple choice questions. The questions are distributed as follows:

Verbal Battery		Quantitative Battery	
Sentence Completion	20	Number Analogies	18
Verbal Classifications	20	Number Series	18
Verbal Analogies	24	Number Puzzles	16
Non-Verbal Battery			
Figure Matrices	22		
Figure Classification	22	Total Questions 176	
Paper Folding	16		

Test Sections

The test consists of verbal, nonverbal material and quantitative material.

VERBAL BATTERY

The verbal battery on the CogAT® is designed to measure a student's vocabulary, memory, ability to solve verbal problems, and ability to determine word relationships.

The verbal battery has three question types.

- ✓ Sentence Completion: Students select the word that best completes the sentence.
- ✓ Verbal Classification: Students are given a series of three words that are in some way similar. The student then selects a word from the answer choices that is connected to the other three.
- ✓ Verbal Analogies: Students are provided with two words that form a pair, as well as a third word. From the answer choices, the student must select the word that goes best with the third provided word.

In the verbal battery, the student must read individual words on two subtests (Verbal Analogies & Verbal Classification) and a sentence on one subtest (Sentence Completion).

NONVERBAL BATTERY

On the nonverbal battery, students are tested on their ability to reason using geometric shapes and figures. Students must use strategies to solve unique problems that they may never have encountered in school.

The nonverbal battery is composed of three question types (subtests):

- ✓ Figure Classification: Students are provided with three figures and must select the fourth figure that completes the set.
- ✓ Figure Matrices: Students are given a 2x2 matrix with the image missing in one cell. Students must determine the relationship between the two spatial forms in the top row and find a fourth image that has the same relationship to the spatial form in the bottom row.
- ✓ Paper Folding: Students must determine how a hole-punched, folded paper will look once it is unfolded.

QUANTITATIVE BATTERY

The quantitative battery measures abstract reasoning, quantitative reasoning, and problem solving skills.

The quantitative battery is composed of three question types (subtests):

- ✓ Number Series: Students are given a series of numbers (terms). Based on the terms in the series, students must determine what the next term in the series should look like.
- ✓ Number Puzzles: Students are asked to solve simple equations by finding a missing value.
- ✓ Number Analogies: Students are provided with two sets of analogous numbers, and a third set with a missing number. To determine the missing number, students must find the relationship between the numbers in each of the first two sets and apply it to the final set.

Part 2: How to Use this Book

The CogAT® is an important test and the more a student is familiar with the questions on the exam, the better she will fare when taking the test.

This book will help your student get used to the format & content of the test so s/he will be adequately prepared and feel confident on test day.

Inside this book, you will find:

* Sample question for each question type on the test and teaching tips to help your child approach each question type strategically and with confidence.

* Full-length CogAT® Level 11 practice test.

* Access to bonus practice questions online at https://originstutoring.lpages.co/cogat-11-challenge-questions/

Part 3. Test Prep Tips and Strategies

Firstly, and most importantly, commit to make the test preparation process a stress-free one. A student's ability to keep calm and focused in the face of challenge is a quality that will benefit her throughout her academic life.

Be prepared for difficult questions from the get-go! There will be a certain percentage of questions that are very challenging for all children. It is key to encourage students to use all strategies available when faced with challenging questions. And remember that a student can get quite a few questions wrong and still do very well on the test.

Before starting the practice test, go through the sample questions and read the teaching tips provided at the beginning of the book. They will help you guide your student as he or she progresses through the practice test.

The following strategies may also be useful as you help your child prepare:

Before You Start

Find a quiet, comfortable spot to work free of distractions. Show your student how to perform the simple technique of shading (and erasing) bubbles.

During Prep

If your student is challenged by a question, ask your child to explain why he or she chose a specific answer. If the answer was incorrect, this will help you identify where your student is stumbling. If the answer was correct, asking your student to articulate her reasoning aloud will help reinforce the concept.

Encourage your student to carefully consider all the answer options before selecting one. Tell him or her there is only ONE answer.

If your student is stumped by a question, she or he can use the process of elimination. First, encourage your student to eliminate obviously wrong answers to narrow down the answer choices. If your student is still in doubt after using this technique, tell him or her to guess as there are no points deducted for wrong answers.

Review all the questions your student answered incorrectly, and explain to your student why the answer is incorrect. Have your student attempt these questions again a few days later to see if he now understands the concept.

Encourage your student to do her best, but take plenty of study breaks. Start with 15-20 minute sessions. Your student will perform best if she views these activities as fun and engaging, not as exercises to be avoided.

When to Start Preparing?

Every parent/teacher & student will approach preparation for this test differently. There is no 'right' way to prepare; there is only the best way for a particular student. We suggest students, at minimum, take one full-length practice test and spend 6-8 hours reviewing CogAT® practice questions.

If you have limited time to prepare, spend most energy reviewing areas where your student is encountering the majority of problems.

As they say, knowledge is power! Preparing for the CogAT® will certainly help your student avoid anxiety and make sure she does not give up too soon when faced with unfamiliar and perplexing questions.

CogAT® Verbal Battery
Sample Questions & Teaching Tips

This battery includes three types of questions (subtests).

i.   Sentence Completion
ii. Verbal Classifications
iii. Verbal Analogies

If your student/child is going to go through the battery as if s/he were taking it under real testing conditions, then you should allow 30 minutes to complete this battery (10 minutes for each subtest).

Otherwise, we suggest that your student answers each question in his or her own time, while you guide, support and give feedback as s/he progresses. In this case, we also recommend that you spend a few minutes yourself reviewing the teaching tips for each section so you can be prepared to help your student if he or she struggles with a question.

Before starting the battery, have your child try the sample questions in the next few pages.

Sentence Completion

There are 20 Sentence Completion questions in the CogAT® Level 11.

SAMPLE QUESTION - Find the word that best completes the sentence.

The _____ of children are ready to take the test. They have all been preparing for weeks and know the material very well.

A. majority   B. most   C. rest   D. minority   E. intelligent

Correct Answer: A. The word 'majority' is the best word as it relates to most of the children who have 'all been preparing for weeks and know the material very well.'

TEACHING TIPS

* Encourage your student to carefully read the sentence in order to understand its meaning and structure before moving on the answer choices.
* Ask your student to try to predict what word could be used in the sentence.
* Encourage your student to identify 'clue' words/phrases. These words can help point a student towards the correct answer. For example, "however" or "but" indicates the second part of the sentence will be a caveat or mitigation, while words like "furthermore" mean further support or elaboration on a point.

Verbal Classifications

There are 20 Verbal Classifications questions in the CogAT® Level 11.

SAMPLE QUESTION: Pick a word from the answer choices that belongs in the same group as the words in the top row.

| basement | attic | kitchen |

A. downstairs   B. roof   C. garden   D. house   E. dining room

Correct Answer: **E.** The similarity among the items is that they are all rooms in a house.

TEACHING TIPS
- These questions test a student's ability to identify and classify common objects into basic categories by one or more common physical property or attribute. They also test knowledge of common objects and categories. Introduce these categories and characteristics in real-life situations and discuss the relationships between concepts with your student.

  - Objects: Tools, stationary, musical instruments, kitchen instruments, land and air transportation objects.
  - Animals: Birds, insects, sea or land animals, animal homes, animal babies
  - Fruits, vegetables and spices
  - Feelings/emotions
  - States, Continents, Countries
  - Professions. Sports
  - Grammatical constructs: nouns, verbs, adjectives, adverbs.
  - Elements in nature. Shapes

- Encourage your student to expand on his knowledge of a category in a question. Ask him to name other objects that share the same characteristics and belong to a specific category.

- Ask your student to explain why she chose a specific answer. This will help you identify where your student is stumbling or provide the opportunity to reinforce understanding of a category and the object/s that can "belong" to it.

Verbal Analogies

There are 24 Verbal Analogies questions in the CogAT® Level 11.

SAMPLE QUESTION: Find the relationship between the first two words, then choose a word that has the same relationship with the third word.

teacher : student   as   doctor :

A. hospital   B. patient   C. passenger   D. medicine   E. customer

Correct Answer: **B.** Teachers help students as doctors help patients.

TEACHING TIPS

- To master analogies, a student needs to have general background knowledge, and an understanding/recognition of various relationships, including:
  - → Object/function — One word in a pair describes the purpose or function of the other word.
  - → Agent (person or animal)/location.
  - → Agent (person or animal)/action.
  - → Definition/Evidence—One word in a pair helps to define the other word; or, one word in a pair is a defining characteristic of the other word.
  - → Synonym/Antonym—One word in a pair is a synonym or antonym of the other word.
  - → Degree/Intensity—Both words in a pair are similar in concept, but vary in intensity.
  - → Component/Part—One word in a pair represents one part of the other word, which represents a whole; or, one word is simply a component of the other.

- As often as possible, incorporate discussions about similarities, differences, and relationships between words into your everyday conversation with your student. Help your student begin thinking about how different words and concepts are connected to one another.

- When answering practice questions, teach your student to determine the relationship between the first pair of words before looking at the answer choices. Once your student determines the relationship between the first pair, she can then look at the answer choices to find the pair with the exact same relationship.

CogAT® Nonverbal Battery
Sample Questions & Teaching Tips

This battery includes three types of questions (subtests).

  i.   Figure Classification
  ii.   Figure Matrices
  iii. Paper Folding

If your student/child is going to go through the battery as if s/he were taking it under real testing conditions, then you should allow 30 minutes to complete this battery (10 minutes for each subtest).

Otherwise, have your student answer each question in his or her own time, while you guide, support and give feedback as she progresses. Spend a few minutes reviewing the teaching tips for each section so you help your student if he struggles with a question.

Before starting the battery, ask your student to try the sample questions in the next few pages.

Figure Classification

There are 22 Figure Classification questions in the CogAT® Level 11.

SAMPLE QUESTION:

Look at the shapes in the top row. These shapes go together in a certain way. Which shape in the bottom row belongs with the shapes in the top row?

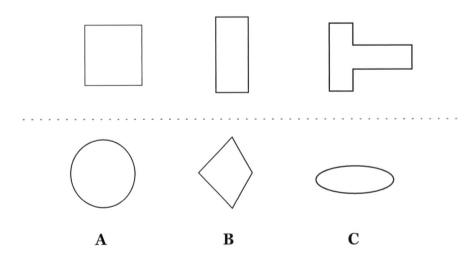

Correct Answer: **B.** In the top row, the three figures go together because all the shapes are made of straight lines. Your student needs to find the figure among the answer options that goes together with the shapes in the top row. Option A and C are incorrect because these shapes do not have straight lines. Option B is correct as this is the only shape that has straight lines like the shapes in the top row.

TEACHING TIPS

- After your student has answered the question, encourage her to expand on her knowledge of the category in the question.

- Ask him to name or draw other objects that share the same characteristics and belong to a specific category.

- Ask your student to explain why she chose a specific answer. This will help you identify where your student is stumbling or provide the opportunity to reinforce understanding of a category and the object/s that can "belong" to it.

Figure Matrices

There are 22 Figure Matrices questions in the CogAT® Level 11.

SAMPLE QUESTION:

Look at the shapes in the boxes on top. These shapes go together in a certain way. Which answer choice belongs where the question mark is?

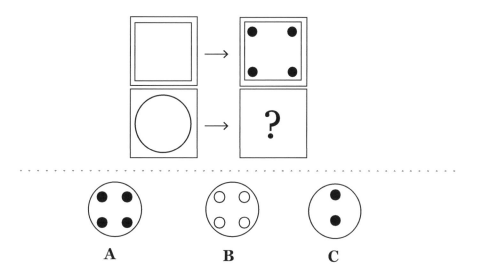

Correct Answer: **A.** In the top row, there are two figures that go together in a certain way. They go together in the sense that as the figure moves from left box to right box, it stays the same shape (a square) but adds four black circles inside.

Your student needs to find the figure among the answer options that fits best in the question mark box on the bottom row. The correct choice will have the same relationship with the figure on the bottom row that the figures in the top row have with each other.

Option B is incorrect because, although the figure is the same shape as the figure on the bottom row, the inside circles that are added are white. Option C is incorrect because, although the figure is the same shape as the figure on the bottom row, only two black inside circles are added. Option A is correct as the figure has the same shape (circle) as the figure on the bottom row and it has four black circles inside.

TEACHING TIPS

- Make sure your student knows key concepts that come up in these types of questions, including geometric concepts such as rotational symmetry, line symmetry, parts of a whole.

- If your student is finding these items difficult, encourage her to discover the pattern by isolating one element (e.g: outer shape, inner shape/s) and identify how it changes:

  → Ask: Is the color/shading of the element changing as it moves?

  → Ask: Is the element changing positions as it moves? Does it move up or down? Clockwise or counter-clockwise? Does it end up in the opposite (mirror) position?

  → Ask: Does the element disappear or increase in number as it moves along the row? Does it get bigger or smaller?

- Encourage your student to make a prediction for the missing object and compare the description with the answer choices.

Paper Folding

There are 16 Paper Folding questions in the CogAT® Level 11.

SAMPLE QUESTION:

The paper in the top row is folded and cut as shown. Which paper in the bottom row is the result when the paper is unfolded?

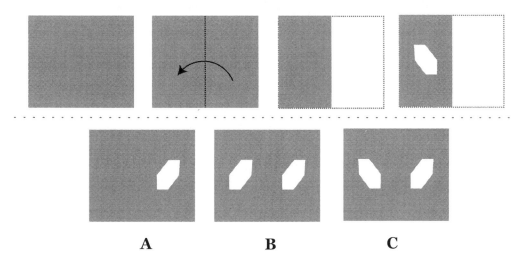

A   **B**   **C**

Correct Answer: **C.** In the top row, the paper is folded in half vertically. A shape is cut out of the folded paper. Your child needs to choose which answer option shows how the paper will look when unfolded. The correct choice will be an unfolded paper with two shapes cut out of it. The shape on the right will mirror the shape on the left. Option A is incorrect as it shows only one shape cut out from the paper. Option B is incorrect as both shapes are placed in the same direction and are not mirror images of each other. Option C is correct as the right shape reflects the left shape, like a mirror image.

TEACHING TIPS

- In addition to using the written practice questions, a good way to prepare for this unique and challenging question type is through hands-on practice with real paper. For example, you can show your student that if a paper is folded once and a hole is punched into it, she will see two holes on either side of the fold once the paper is unfolded.

CogAT®Quantitative Battery
Sample Questions & Teaching Tips

This battery includes three types of questions (subtests).

i. Number Series
ii. Number Puzzles
iii. Number Analogies

If your student/child is going to go through the battery as if s/he were taking it under real testing conditions, then you should allow 30 minutes to complete this battery (10 minutes for each subtest).

Otherwise, have your student answer each question in his or her own time, while you guide and give feedback as s/he progresses. Spend a few minutes reviewing the teaching tips for each section so you can help your student if he struggles with a question.

Before starting the battery, have your student try the sample questions in the next few pages.

Number Series

There are 18 Number Series questions in the CogAT® Level 11.

SAMPLE QUESTION: Pick a number which follows the same rule as the order of numbers in the top row and replace the question mark with this number.

40	34	28	22	16	10	?

A. 4	B. 6	C. 16	D. 8	E. 2

Correct Answer: **A.**   Subtract 6 from each term in the series.

TEACHING TIPS

• Because your student may not have formal academic experience with this question type. it is important to practice working with many of these questions before the test. You can also find workbooks or games related to number patterns and sequences to help your child further his or her understanding of these concepts in an engaging manner.

Number Puzzles

There are 16 Number Puzzles questions in the CogAT® Level 11. An example of each of the two types of Number Puzzle questions are below:

SAMPLE QUESTION 1: Replace the question mark with the correct number to solve the equation.

$$20 + ? = 5 \times 4$$

A. 5   B. 25   C. 0   D. 9   E. 20

Correct Answer: **C.** $20 + 0 = 5 \times 4$

SAMPLE QUESTION 2: Replace the question mark with the correct number to solve the equation.

$$15 + ? = \bullet$$

$$\bullet = 26$$

A. 26   B. 9   C. 11   D. 13   E. 15

Correct Answer: **C.** $15 + 11 = 26$

TEACHING TIPS

- This question type requires your child to solve basic math equations, so practice with numbers and problem solving is essential.

- Make sure your child understands the meaning of "equal," since the object is to supply the missing piece of information that will make two provided equations equal to one another.

- You can also teach your child to approach the question by "plugging in" the answer choices and solving to see if the result is equal to the other equation in the question.

Number Analogies

There are 18 Number Analogies questions in the CogAT® Level 11.

SAMPLE QUESTION: Find the relationship between the numbers in the first set, and between the numbers in the second set. Then choose a number which follows the same pattern when paired with the number in the third set.

[10 → 2]   [20 → 4]   [30 → ?]

A. 3   **B. 7**   **C. 6**   **D. 4**   **E. 10**

Correct Answer: C. The rule is to divide the first number in each set by 5, so the answer is 6 (option C).

TEACHING TIPS

- Your child is probably not accustomed to completing number matrices, so it is important to frequently expose him to this question type in order to build confidence and familiarity.

- Consider modeling how to approach solving a number matrix by "thinking aloud" as you work through a question with your child.

- Work with your child on basic mathematical concepts, including addition, subtraction, division, multiplication.

COGAT® LEVEL 11
PRACTICE TEST

VERBAL BATTERY

• • • • • • • • • • • • •

SENTENCE COMPLETION PRACTICE QUESTIONS

• • • • • • • • • • • • • • • • • • • •

For each item, the student is presented with a sentence that has a missing word.

The student needs to choose a word from the answer choices that fits best in the empty space to complete the sentence.

SENTENCE COMPLETION

1. **A singer uses a_____ to project her voice around a room.**

   A. loud speaker **B.** microphone **C.** sound **D.** karaoke **E.** song

2. **The politician does not approve of the law and is trying to_____it.**

   A. abolish **B.** enact **C.** legislate **D.** approve **E.** attack

3. **Some cities are very polluted and many people don't want to live in them and_____the air.**

   A. consume **B.** overwhelm **C.** inhale **D.** clean **E.** estimate

4. **There are_____reasons to exercise each day, but one is that you feel better after doing it.**

   A. enough **B.** insufficient **C.** incredible **D.** terrible **E.** numerous

5. **A farmer uses a_____as one of his tools.**

   A. pitchfork **B.** farm **C.** car **D.** scalpel **E.** cow

6. **The tyrannosaurus is_____like all other dinosaurs**

   A. threatened   **B.** extinct   **C.** disadvantaged   **D.** fierce   **E.** abundant

7. **The ship_____before it sank.**

   A. capsized   **B.** swerved   **C.** drowned   **D.** captivated   **E.** jumped

8. **The students'_____consisted of 10 bunkbeds.**

   A. laboratory   **B.** library   **C.** room   **D.** list   **E.** dormitory

9. **Instead of being boastful of his wealth and power, the king was quite_____about his good fortune**

   A. proud   **B.** sorrowful   **C.** rich   **D.** humble   **E.** greedy

10. **Sarah said she did not have enough money so she could not_____a concert ticket**

   A. save   **B.** afford   **C.** sell   **D.** price   **E.** have

SENTENCE COMPLETION

11. **Snakes are a popular pet, despite the fact they can be_____.**

    A. unpopular   **B.** favored   **C.** dangerous   **D.** widespread   **E.** fun

12. **Don't estimate", said John. "We need _____ number".**

    A. an accurate   **B.** a true   **C.** a large   **D.** a perfect   **E.** an imprecise

13. **The principal decided to _____ the student for his misdeeds.**

    A. praise   **B.** move   **C.** expel   **D.** deport   **E.** exile

14. **The_____of the church could be seen from miles away.**

    A. priest   **B.** congregant   **C.** alter   **D.** doorknob   **E.** steeple

15. **Parrots usually live wild in the _____, but some people have them as pets.**

    A. zoo   **B.** tropics   **C.** desert   **D.** humidity   **E.** heat

SENTENCE COMPLETION

16. Everyone wanted to be her friend; they enjoyed her kind demeanor and _____ attitude.

   A. dark   **B.** upbeat   **C.** popular   **D.** demanding   **E.** active

17. "I don't agree," said Scott. "I have a different _____.

   A. prospect   **B.** question   **C.** dimension   **D.** subjective   **E.** perspective

18. They were arrested and held in custody and are _____ trial.

   A. passing   **B.** abusing   **C.** avoiding   **D.** awaiting   **E.** anointing

19. Wash your hands so you are less likely to be _____ to the virus.

   A. caught   **B.** spread   **C.** infected   **D.** exposed   **E.** examined

20. "You need to _____ your words better," said the drama teacher, "I can barely understand you."

   A. articulate   **B.** scream   **C.** offer   **D.** anticipate   **E.** announce

VERBAL BATTERY

· · · · · · · · · · ·

VERBAL CLASSIFICATIONS PRACTICE QUESTIONS

· · · · · · · · · · · · · · · · · · ·

For each item, the student is presented with three words on the top row. The students needs to figure out how these words are related.

The student then needs to pick a word from the answer choices that belongs in the same group as the three words in the top row.

VERBAL CLASSIFICATIONS

1. **opossum   kangaroo   koala**

    A. deer   **B.** wallaby   **C.** Australia   **D.** animal   **E.** cute

2. **boxing   skiing   surfing**

    A. judo   **B.** basketball   **C.** games   **D.** baseball   **E.** playing

3. **calf   ankle   knee**

    A. shoulder   **B.** kick   **C.** cow   **D.** finger   **E.** thigh

4. **incisor   bicuspid   molar**

    A. canine   **B.** bite   **C.** teeth   **D.** tongue   **E.** cavity

5. **iris   pupil   retina**

    A. eye   **B.** cornea   **C.** temple   **D.** color   **E.** see

VERBAL CLASSIFICATIONS

6. **louse   tick   tapeworm**

    A. parasite   **B.** dirty   **C.** ant   **D.** flea   **E.** earthworm

7. **beret   fedora   turban**

    A. head   **B.** cap   **C.** clothes   **D.** hats   **E.** scarf

8. **doctor   dentist   nurse**

    A. medicine   **B.** hospital   **C.** waitress   **D.** vet   **E.** librarian

9. **century   decade   minute**

    A. time   **B.** measure   **C.** centimeter   **D.** month   **E.** mile

10. **silo   barn   pen**

    A. coop   **B.** farm   **C.** pigs   **D.** hay   **E.** enclosure

11. **announce   declare   publicize**

   A. hide   **B.** suppress   **C.** broadcast   **D.** commercial   **E.** advertisement

12. **Sahara   Gobi   Great Sandy**

   A. desert   **B.** Mojave   **C.** sand   **D.** London   **E.** Andes

13. **Himalayas   Rocky Mountains   Alps**

   A. mountains   **B.** Andes   **C.** Antarctica   **D.** high   **E.** Pluto

14. **waiter   florist   chef**

   A. job   **B.** plumber   **C.** restaurant   **D.** party   **E.** diner

15. **mare   ewe   hen**

   A. lion   **B.** stallion   **C.** animals   **D.** lioness   **E.** farm

16. **calm   peaceful   tranquil**

    A. troubled   **B.** serene   **C.** ocean   **D.** state   **E.** opaque

17. **dragon   unicorn   werewolf**

    A. mermaid   **B.** mythical   **C.** Greek   **D.** anteater   **E.** creature

18. **Muslim   Christian   Jewish**

    A. religion   **B.** church   **C.** hispanic   **D.** Hindu   **E.** Indian

19. **pillow   sheet   mattress**

    A. bedroom   **B.** sleep   **C.** comforter   **D.** cozy   **E.** lullaby

20. **cinnamon   cardamom   cloves**

    A. cumin   **B.** coffee   **C.** cocoa   **D.** cream   **E.** spice

VERBAL BATTERY

· · · · · · · · · ·

VERBAL ANALOGIES PRACTICE QUESTIONS

· ·

For each item, the student is presented with two words that have a relationship or go together in a particular way.

The student needs to figure out the relationship between those first two words. The student then needs to choose the word in the answer choices that has the same relationship with the third word.

VERBAL ANALOGIES

1. **boxer : gloves as carpenter : ?**

   A. craft   **B.** tools   **C.** ant   **D.** fighter   **E.** table

2. **frog: amphibian as koala : ?**

   A. tadpole   **B.** pouch   **C.** marsupial   **D.** tree   **E.** reptile

3. **skiing : slope as driving : ?**

   A. car   **B.** glide   **C.** road   **D.** traffic   **E.** cross country

4. **petite : tiny as obese : ?**

   A. slender   **B.** large   **C.** eating   **D.** person   **E.** small

5. **saturn: planet as incisor : ?**

   A. measure   **B.** molar   **C.** mouth   **D.** solar   **E.** tooth

VERBAL ANALOGIES

6. **turkey : bird as cactus : ?**

   A. prickly   **B.** plant   **C.** desert   **D.** seaweed   **E.** water

7. **cyclist : track as swimmer: ?**

   A. race   **B.** driver   **C.** pool   **D.** goggles   **E.** swim

8. **hair stylist: salon as pilot : ?**

   A. cabin   **B.** scissors   **C.** wings   **D.** cockpit   **E.** ship

9. **librarian : library as bellhop: ?**

   A. books   **B.** work   **C.** profession   **D.** hotel   **E.** luggage

10. **consent : oppose as primitive : ?**

    A. surrender   **B.** infant   **C.** young   **D.** ancient   **E.** modern

VERBAL ANALOGIES

11. **bright : brilliant as cry : ?**

   A. shine   **B.** tear   **C.** sob   **D.** smile   **E.** speak

12. **vacant : empty as pupil : ?**

   A. school   **B.** eye   **C.** teacher   **D.** unoccupied   **E.** student

13. **lion : cage as valuables : ?**

   A. money   **B.** jewelry   **C.** box   **D.** vault   **E.** zoo

14. **gas tank : gas as bone : ?**

   A. car   **B.** marrow   **C.** flesh   **D.** leg   **E.** boney

15. **provoke : soothe as loathe : ?**

   A. cherish   **B.** aggravate   **C.** detest   **D.** insult   **E.** win

16. **feet : yard as quart : ?**

   A. gallon   **B.** half   **C.** inch   **D.** kilometer   **E.** measurement

17. **debate : agree as excavate : ?**

   A. dig   **B.** dispute   **C.** consent   **D.** bury   **E.** scrape

18. **pencil : graphite as thermometer : ?**

   A. temperature   **B.** sick   **C.** mercury   **D.** lead   **E.** iodine

19. **rural : soil as urban : ?**

   A. city   **B.** skyscraper   **C.** pavement   **D.** sand   **E.** dirt

20. **breeze : gale as drip : ?**

   A. cascade   **B.** wash   **C.** storm   **D.** storm   **E.** fall

VERBAL ANALOGIES

21. **pack : wolf as pod : ?**

    A. school   **B.** whale   **C.** fish   **D.** pea   **E.** class

22. **desk : wood as tire : ?**

    A. material   **B.** rubber   **C.** car   **D.** plastic   **E.** drive

23. **kangaroo : joey as bird : ?**

    A. blackbird   **B.** eaglet   **C.** fly   **D.** baby   **E.** fledgling

24. **absurd : sensible as fatigue : ?**

    A. tired   **B.** crazy   **C.** liveliness   **D.** clumsy   **E.** exhausted

NON-VERBAL BATTERY

· · · · · · · · · · · ·

FIGURE CLASSIFICATIONS PRACTICE QUESTIONS

· ·

Figure Classifications

Look at the shapes in the top row. These shapes go together in a certain way. Which shape in the bottom row belongs with the shapes in the top row?

1

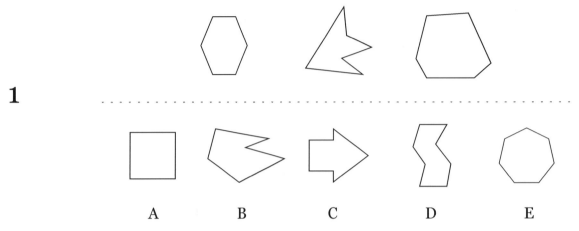

A	B	C	D	E

Look at the shapes in the top row. These shapes go together in a certain way. Which shape in the bottom row belongs with the shapes in the top row?

2

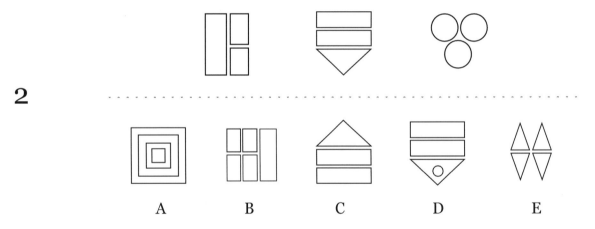

A	B	C	D	E

Look at the shapes in the top row. These shapes go together in a certain way. Which shape in the bottom row belongs with the shapes in the top row?

3

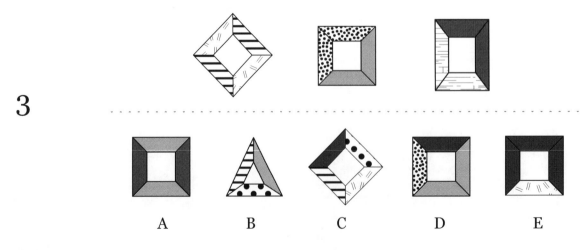

A	B	C	D	E

Look at the shapes in the top row. These shapes go together in a certain way. Which shape in the bottom row belongs with the shapes in the top row?

4

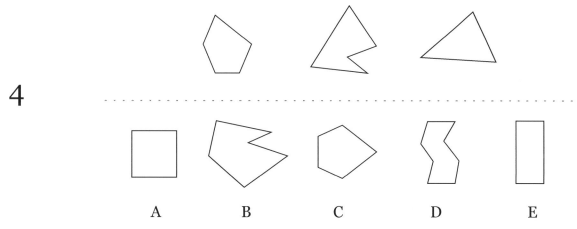

Look at the shapes in the top row. These shapes go together in a certain way. Which shape in the bottom row belongs with the shapes in the top row?

5

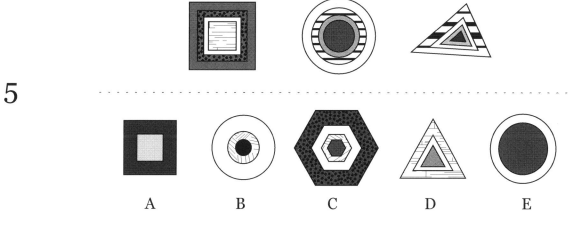

Look at the shapes in the top row. These shapes go together in a certain way. Which shape in the bottom row belongs with the shapes in the top row?

6

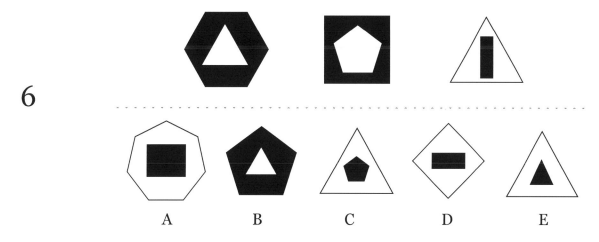

Figure Classifications

Look at the shapes in the top row. These shapes go together in a certain way. Which shape in the bottom row belongs with the shapes in the top row?

7

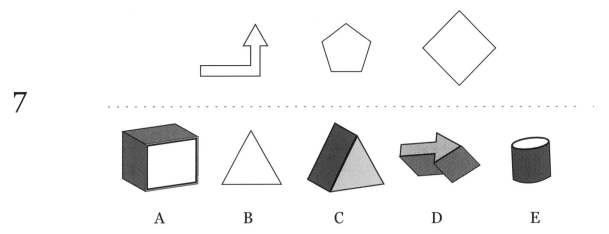

Look at the shapes in the top row. These shapes go together in a certain way. Which shape in the bottom row belongs with the shapes in the top row?

8

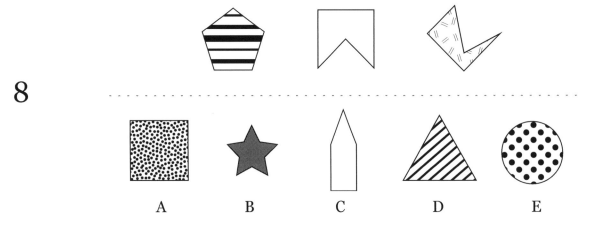

Look at the shapes in the top row. These shapes go together in a certain way. Which shape in the bottom row belongs with the shapes in the top row?

9

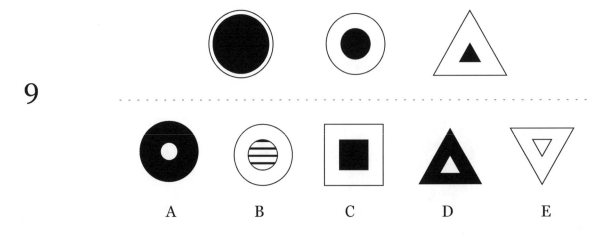

CogAT® Level 11 Test Prep Book

Figure Classifications

Look at the shapes in the top row. These shapes go together in a certain way. Which shape in the bottom row belongs with the shapes in the top row?

10

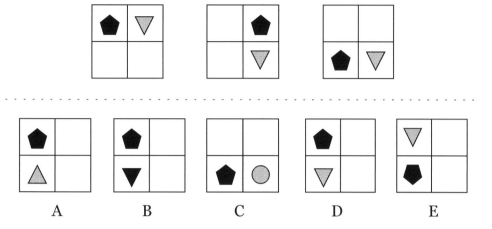

A   B   C   D   E

Look at the shapes in the top row. These shapes go together in a certain way. Which shape in the bottom row belongs with the shapes in the top row?

11

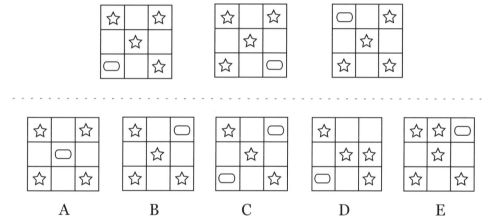

A   B   C   D   E

Look at the shapes in the top row. These shapes go together in a certain way. Which shape in the bottom row belongs with the shapes in the top row?

12

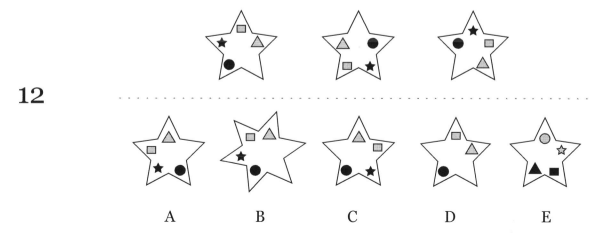

A   B   C   D   E

Figure Classifications

Look at the shapes in the top row. These shapes go together in a certain way. Which shape in the bottom row belongs with the shapes in the top row?

13

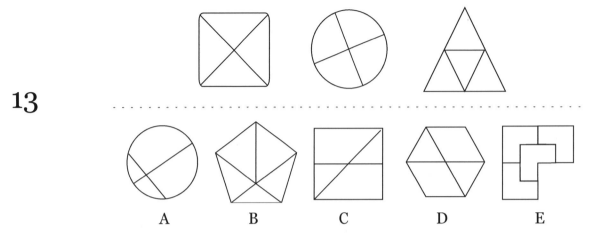

Look at the shapes in the top row. These shapes go together in a certain way. Which shape in the bottom row belongs with the shapes in the top row?

14

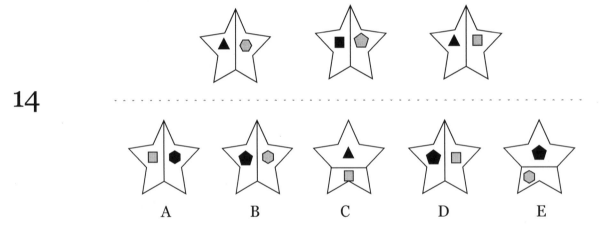

Look at the shapes in the top row. These shapes go together in a certain way. Which shape in the bottom row belongs with the shapes in the top row?

15

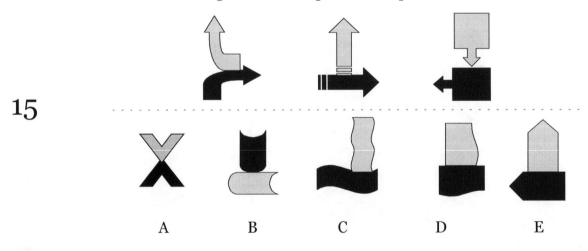

Figure Classifications

Look at the shapes in the top row. These shapes go together in a certain way. Which shape in the bottom row belongs with the shapes in the top row?

16

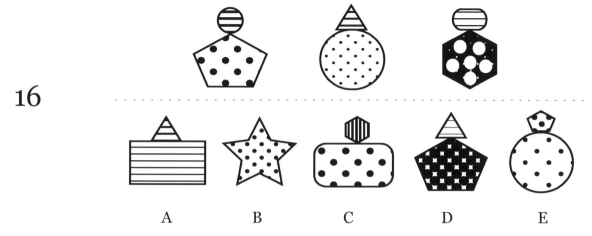

A   B   C   D   E

Look at the shapes in the top row. These shapes go together in a certain way. Which shape in the bottom row belongs with the shapes in the top row?

17

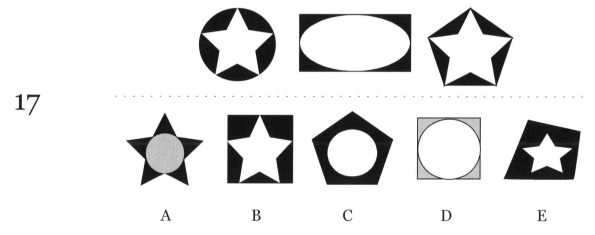

A   B   C   D   E

Look at the shapes in the top row. These shapes go together in a certain way. Which shape in the bottom row belongs with the shapes in the top row?

18

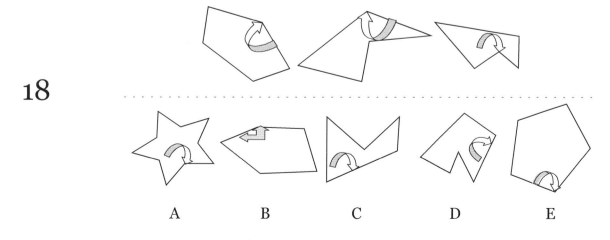

A   B   C   D   E

Figure Classifications

Look at the shapes in the top row. These shapes go together in a certain way. Which shape in the bottom row belongs with the shapes in the top row?

19

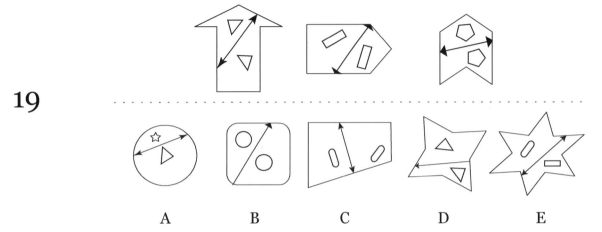

| A | B | C | D | E |

Look at the shapes in the top row. These shapes go together in a certain way. Which shape in the bottom row belongs with the shapes in the top row?

20

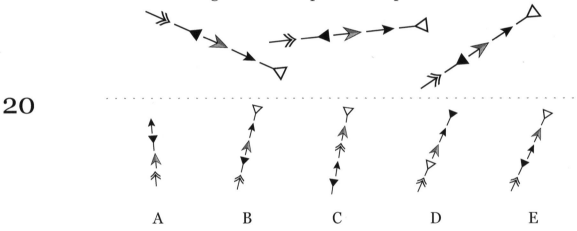

| A | B | C | D | E |

Look at the shapes in the top row. These shapes go together in a certain way. Which shape in the bottom row belongs with the shapes in the top row?

21

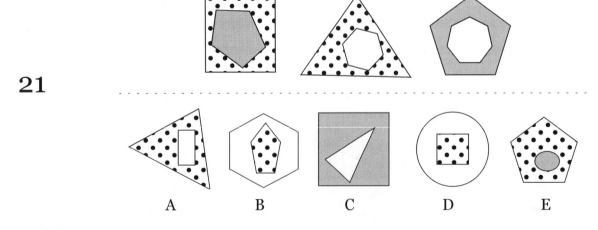

| A | B | C | D | E |

Look at the shapes in the top row. These shapes go together in a certain way. Which shape in the bottom row belongs with the shapes in the top row?

22

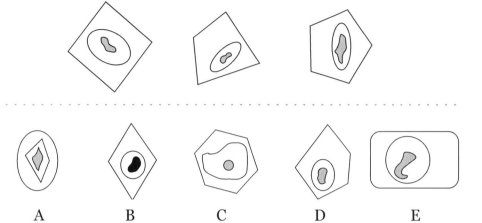

A   B   C   D   E

NON-VERBAL BATTERY

•••••••••••••••

FIGURE MATRICES

PRACTICE QUESTIONS

••••••••••••••••••••••••

Figure Matrices

Look at the shapes in the boxes on top. These shapes go together in a certain way. Which shape belongs where the question mark is?

1

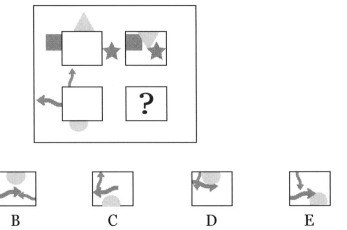

| A | B | C | D | E |

Look at the shapes in the boxes on top. These shapes go together in a certain way. Which shape belongs where the question mark is?

2

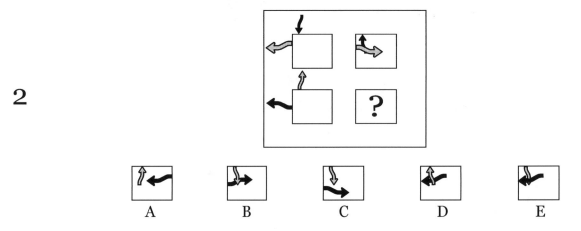

| A | B | C | D | E |

Look at the shapes in the boxes on top. These shapes go together in a certain way. Which shape belongs where the question mark is?

3

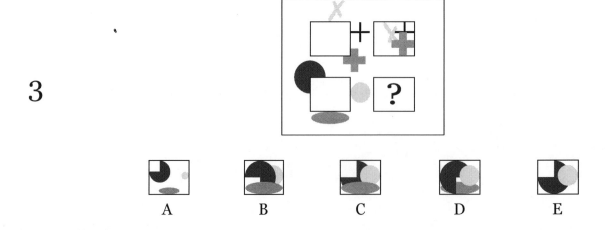

| A | B | C | D | E |

Look at the shapes in the boxes on top. These shapes go together in a certain way. Which shape belongs where the question mark is?

4

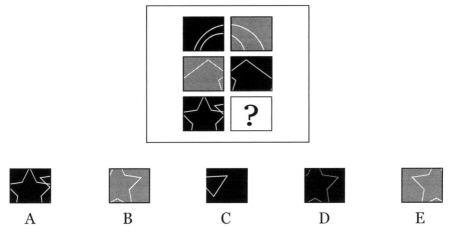

Look at the shapes in the boxes on top. These shapes go together in a certain way. Which shape belongs where the question mark is?

5

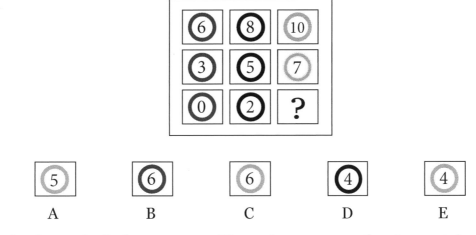

Look at the shapes in the boxes on top. These shapes go together in a certain way. Which shape belongs where the question mark is?

6

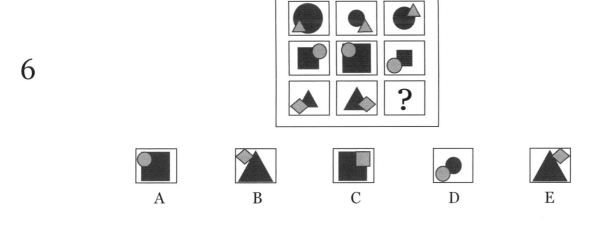

Figure Matrices

Look at the shapes in the boxes on top. These shapes go together in a certain way. Which shape belongs where the question mark is?

7

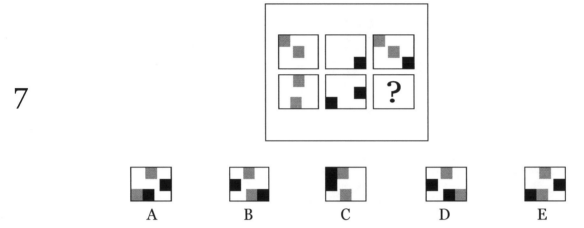

Look at the shapes in the boxes on top. These shapes go together in a certain way. Which shape belongs where the question mark is?

8

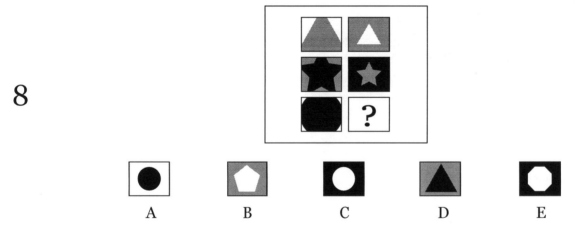

Look at the shapes in the boxes on top. These shapes go together in a certain way. Which shape belongs where the question mark is?

9

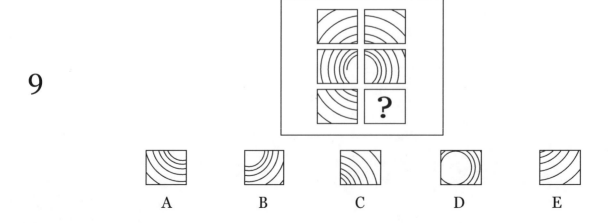

Look at the shapes in the boxes on top. These shapes go together in a certain way. Which shape belongs where the question mark is?

10

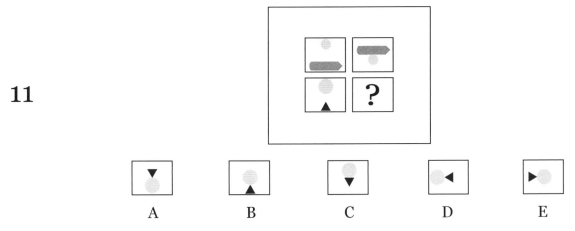

Look at the shapes in the boxes on top. These shapes go together in a certain way. Which shape belongs where the question mark is?

11

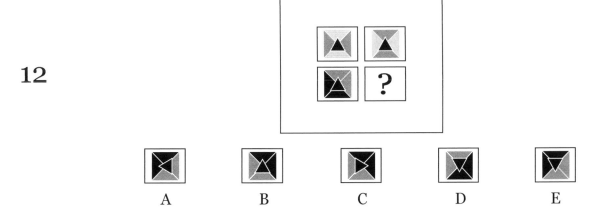

Look at the shapes in the boxes on top. These shapes go together in a certain way. Which shape belongs where the question mark is?

12

Figure Matrices

Look at the shapes in the boxes on top. These shapes go together in a certain way. Which shape belongs where the question mark is?

13

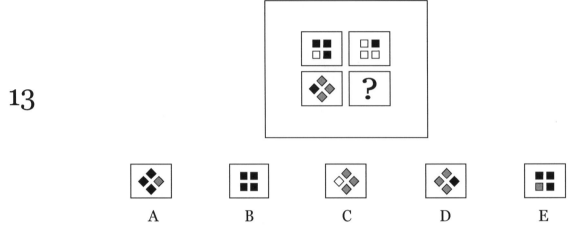

Look at the shapes in the boxes on top. These shapes go together in a certain way. Which shape belongs where the question mark is?

14

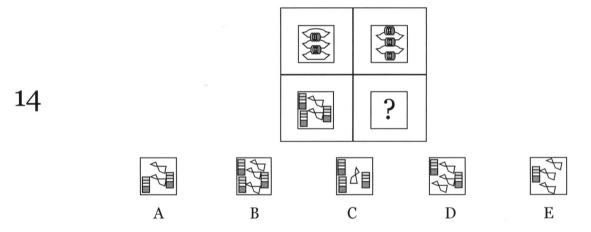

Look at the shapes in the boxes on top. These shapes go together in a certain way. Which shape belongs where the question mark is?

15

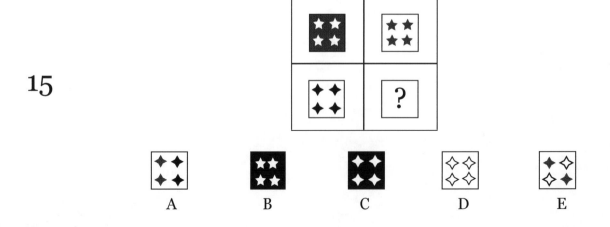

Figure Matrices

Look at the shapes in the boxes on top. These shapes go together in a certain way. Which shape belongs where the question mark is?

16

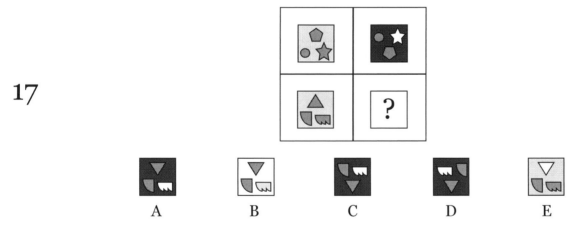

Look at the shapes in the boxes on top. These shapes go together in a certain way. Which shape belongs where the question mark is?

17

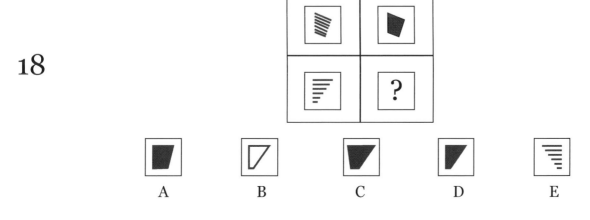

Look at the shapes in the boxes on top. These shapes go together in a certain way. Which shape belongs where the question mark is?

18

Figure Matrices

Look at the shapes in the boxes on top. These shapes go together in a certain way. Which shape belongs where the question mark is?

19

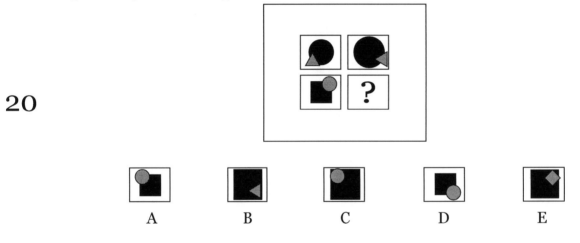

Look at the shapes in the boxes on top. These shapes go together in a certain way. Which shape belongs where the question mark is?

20

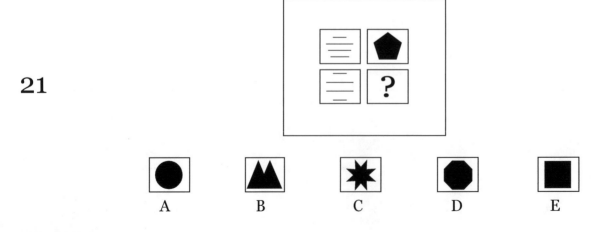

Look at the shapes in the boxes on top. These shapes go together in a certain way. Which shape belongs where the question mark is?

21

Look at the shapes in the boxes on top. These shapes go together in a certain way. Which shape belongs where the question mark is?

22

A	B	C	D	E

NON-VERBAL BATTERY

• • • • • • • • • • •

PAPER FOLDING PRACTICE QUESTIONS

• •

Paper Folding

The paper in the top row is folded and cut as shown. Which paper in the bottom row is the result when the paper is unfolded?

1

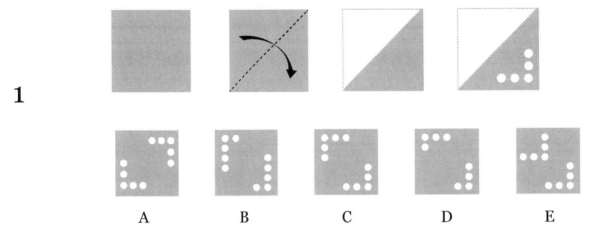

A   B   C   D   E

The paper in the top row is folded and cut as shown. Which paper in the bottom row is the result when the paper is unfolded?

2

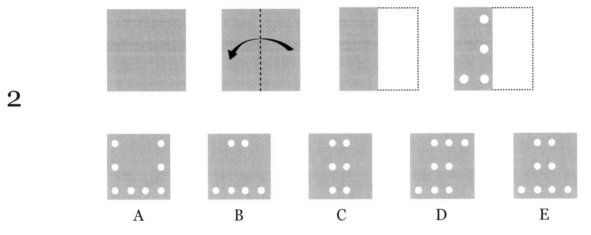

A   B   C   D   E

The paper in the top row is folded and cut as shown. Which paper in the bottom row is the result when the paper is unfolded?

3

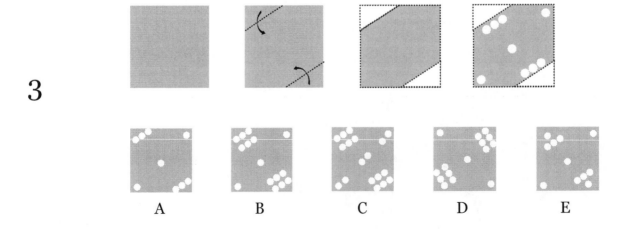

A   B   C   D   E

CogAT® Level 11 Test Prep Book   Gifted & Talented Test Prep Team

The paper in the top row is folded and cut as shown. Which paper in the bottom row is the result when the paper is unfolded?

4

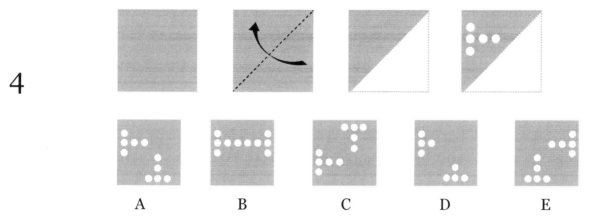

The paper in the top row is folded and cut as shown. Which paper in the bottom row is the result when the paper is unfolded?

5

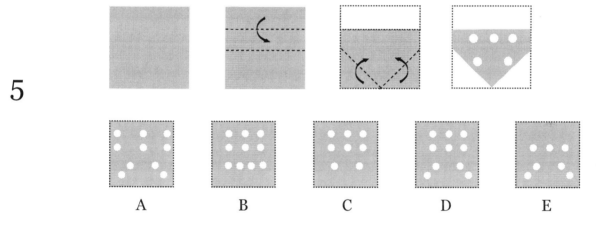

The paper in the top row is folded and cut as shown. Which paper in the bottom row is the result when the paper is unfolded?

6

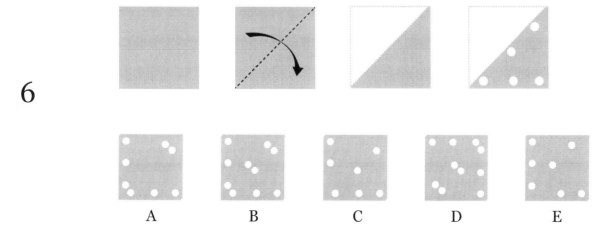

Paper Folding

The paper in the top row is folded and cut as shown. Which paper in the bottom row is the result when the paper is unfolded?

7

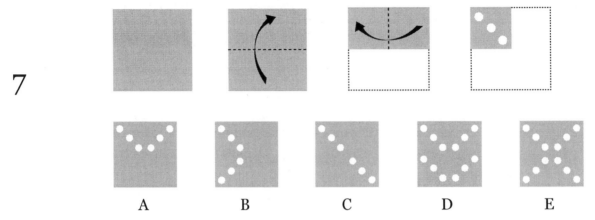

A   B   C   D   E

The paper in the top row is folded and cut as shown. Which paper in the bottom row is the result when the paper is unfolded?

8

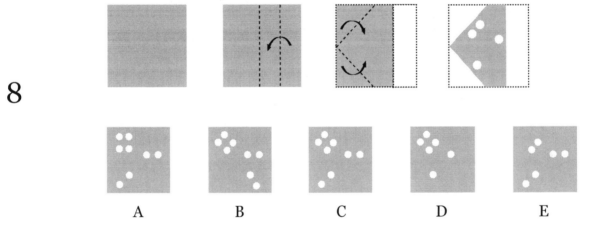

A   B   C   D   E

The paper in the top row is folded and cut as shown. Which paper in the bottom row is the result when the paper is unfolded?

9

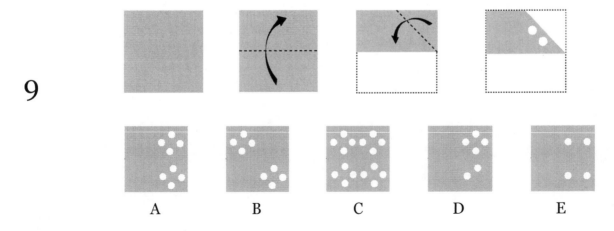

A   B   C   D   E

CogAT® Level 11 Test Prep Book   Gifted & Talented Test Prep Team

Paper Folding

The paper in the top row is folded and cut as shown. Which paper in the bottom row is the result when the paper is unfolded?

10

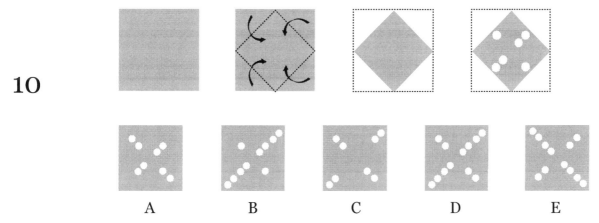

| A | B | C | D | E |

The paper in the top row is folded and cut as shown. Which paper in the bottom row is the result when the paper is unfolded?

11

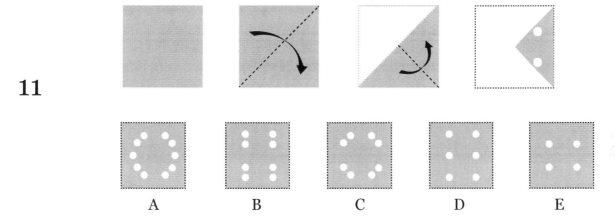

| A | B | C | D | E |

The paper in the top row is folded and cut as shown. Which paper in the bottom row is the result when the paper is unfolded?

12

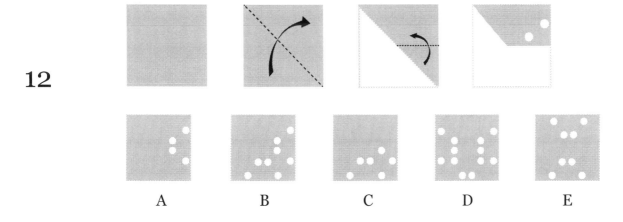

| A | B | C | D | E |

Paper Folding

The paper in the top row is folded and cut as shown. Which paper in the bottom row is the result when the paper is unfolded?

13

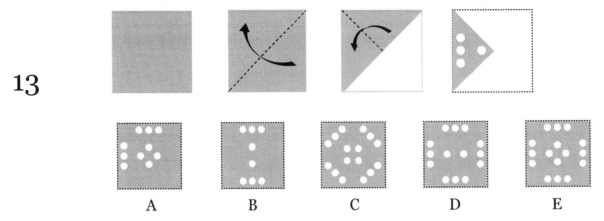

| A | B | C | D | E |

The paper in the top row is folded and cut as shown. Which paper in the bottom row is the result when the paper is unfolded?

14

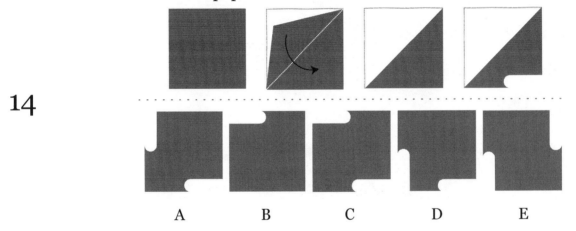

| A | B | C | D | E |

The paper in the top row is folded and cut as shown. Which paper in the bottom row is the result when the paper is unfolded?

15

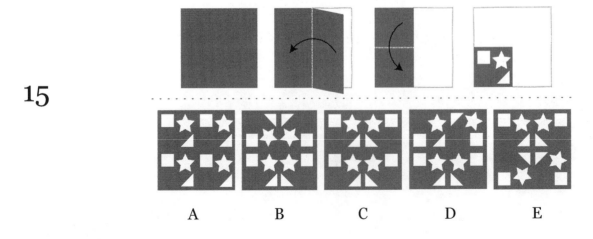

| A | B | C | D | E |

The paper in the top row is folded and cut as shown. Which paper in the bottom row is the result when the paper is unfolded?

16

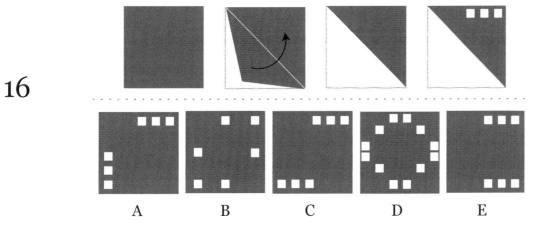

A   B   C   D   E

QUANTITATIVE BATTERY

• • • • • • • • • • •

NUMBER ANALOGIES PRACTICE QUESTIONS

• • • • • • • • • • • • • • • • • • •

For each of the following items, the student is presented with two sets of numbers.

The student needs to find the relationship between those numbers and then choose a number from the answer choices which follows the same pattern when paired with the number in the third set.

NUMBER ANALOGIES

1. [7 ➡ 16]   [15 ➡ 24]   [25 ➡ ?]

   A. 7   **B.** 9   **C.** 34   **D.** 16   **E.** 24

2. [6 ➡ 36]   [8 ➡ 48]   [11 ➡ ?]

   A. 18   **B.** 58   **C.** 51   **D.** 66   **E.** 88

3. [9 ➡ 4 ½]   [6 ➡ 3]   [3 ➡ ?]

   A. 1 ½   **B.** 3   **C.** 2 ½   **D.** ½   **E.** 0

4. [50 ➡ 45]   [5 ➡ 0]   [30 ➡ ?]

   A. 45   **B.** 0   **C.** 15   **D.** 20   **E.** 25

5. [10 ➡ 100]   [100 ➡ 1000]   [50 ➡ ?]

   A. 5000   **B.** 100   **C.** 500   **D.** 1000   **E.** 50,000

6. [7 ➡ 49]   [10 ➡ 100]   [5 ➡ ?]

    A. 30   B. 35   C. 50   D. 25   E. 250

7. [4 ➡ 1]   [16 ➡ 4]   [8 ➡ ?]

    A. 1   B. 2   C. 3   D. 4   E. 6

8. [5 ➡ 5]   [10 ➡ 10]   [0 ➡ ?]

    A. 0   B. 5   C. 10   D. 100   E. 1

9. [2 ➡ 4]   [4 ➡ 16]   [6 ➡ ?]

    A. 25   B. 36   C. 8   D. 18   E. 42

10. [2 ➡ 14]   [4 ➡ 28]   [7 ➡ ?]

    A. 12   B. 22   C. 7   D. 49   E. 56

NUMBER ANALOGIES

11. [1 ➡ 1 ¾]   [1 ¾ ➡ 2 ½]   [2 ½ ➡ ?]

 A. 2 ¾   **B.** 3 ¼   **C.** 4 ¼   **D.** 3 ½   **E.** 3

12. [8 ➡ 13]   [4 ➡ 9]   [10 ➡ ?]

 A. 14   **B.** 18   **C.** 12   **D.** 17   **E.** 15

13. [½ ➡ ¼]   [¼ ➡ 0]   [1 ➡ ?]

 A. ¼   **B.** ½   **C.** ¾   **D.** 1 ¼   **E.** 0

14. [1 ➡ 100]   [10 ➡ 1000]   [0 ➡ ?]

 A. 1000   **B.** 10   **C.** 0   **D.** 1   **E.** 100

15. [2/4 ➡ ½]   [3/6 ➡ ½]   [4/8 ➡ ?]

 A. ½   **B.** ⅓   **C.** 2/8   **D.** 2/6   **E.** ¼

16. [40 ➡ 28]   [30 ➡ 18]   [20 ➡ ?]

    A. 8   **B.** 12   **C.** 18   **D.** 28   **E.** 7

17. [36 ➡ 12]   [30 ➡ 10]   [6 ➡ ?]

    A. 4   **B.** 2   **C.** 4   **D.** 10   **E.** 6

18. [0.01 ➡ 0.1]   [0.002 ➡ 0.02]   [0.3 ➡ ?]

    A. 0.003   **B.** 0.03   **C.** 3   **D.** 0.4   **E.** 30

QUANTITATIVE BATTERY

· · · · · · · · · · ·

NUMBER SERIES
PRACTICE QUESTIONS

· · · · · · · · · · · · · · · · · ·

For each of the following items, the student is presented with a series of numbers. The order of the numbers follow a rule.

The student needs to choose a number from the answer choices which follows the same rule.

NUMBER SERIES

1. **2**　　**8**　　**4**　　**16**　　**8**　　**32**　　**16**　　**?**

　　A. 12　　　　　**B.** 54　　　　　**C.** 48　　　　　**D.** 8　　　　　**E.** 64

2. **12**　　**9**　　**13**　　**8**　　**14**　　**7**　　**15**　　**?**

　　A. 6　　　　　**B.** 16　　　　　**C.** 12　　　　　**D.** 13　　　　　**E.** 10

3. **20**　　**27**　　**23**　　**20**　　**26**　　**13**　　**29**　　**?**

　　A. 30　　　　　**B.** 10　　　　　**C.** 33　　　　　**D.** 7　　　　　**E.** 6

4. **17**　　**20**　　**20**　　**24**　　**24**　　**28**　　**29**　　**?**

　　A. 30　　　　　**B.** 32　　　　　**C.** 28　　　　　**D.** 24　　　　　**E.** 34

5. **21**　　**32**　　**29**　　**40**　　**37**　　**48**　　**45**　　**?**

　　A. 30　　　　　**B.** 36　　　　　**C.** 42　　　　　**D.** 56　　　　　**E.** 58

6. **0.12   0.01   0.10   0.03   0.08   0.05   0.06   [?]**

   A. 0.09   **B.** 0.06   **C.** 0.07   **D.** 0.10   **E.** 0.04

7. **71   73   75   77   76   75   74   76   78   80   [?]**

   A. 80   **B.** 78   **C.** 82   **D.** 81   **E.** 79

8. **55   58   62   67   73   80   [?]**

   A. 83   **B.** 88   **C.** 81   **D.** 89   **E.** 90

9. **23   20   17   22   19   16   21   [?]**

   A. 18   **B.** 24   **C.** 26   **D.** 16   **E.** 27

10. **65   78   91   104   117   [?]**

   A. 120   **B.** 113   **C.** 150   **D.** 130   **E.** 135

11. **0.01**   **0.05**   **0.02**   **0.1**   **0.03**   **0.15**   **0.04**   ☐

    A. 0.01   **B.** 0.1   **C.** 0.2   **D.** 0.05   **E.** 0.02

12. **45**   **41**   **53**   **49**   **61**   **57**   **69**   ☐

    A. 67   **B.** 73   **C.** 66   **D.** 81   **E.** 65

13. **69**   **62**   **53**   **46**   **37**   **30**   ☐

    A. 23   **B.** 25   **C.** 13   **D.** 20   **E.** 21

14. **5**   **7**   **10**   **12**   **15**   **17**   **20**   ☐

    A. 25   **B.** 22   **C.** 23   **D.** 27   **E.** 30

15. **13**   **16**   **20**   **25**   **31**   **38**   **46**   **55**   ☐

    A. 65   **B.** 77   **C.** 110   **D.** 61   **E.** 58

16. **59   58   56   53   49   44   38**   ⬚

   A. 27   **B.** 30   **C.** 31   **D.** 33   **E.** 37

17. 77   75   **80**   78   **83**   81   ⬚

   A. 85   **B.** 79   **C.** 76   **D.** 86   **E.** 83

18. **4**   **12**   **16**   **5**   **15**   **20**   **6**   **18**   ⬚

   A. 14   **B.** 7   **C.** 20   **D.** 24   **E.** 21

QUANTITATIVE BATTERY

• • • • • • • • • • • •

NUMBER PUZZLES PRACTICE QUESTIONS

• • • • • • • • • • • • • • • • • •

For each item, the student is presented with one or more equations.

The student needs to solve the equation and choose a number from the answer choices that will replace the question mark.

NUMBER PUZZLES

1. **?** = ◇ + **13**   ◇ = **12**

   A. 12   **B. 13**   **C.** 23   **D.** 22   **E.** 25

2. **? + 5 = 4 x** ◇   **5 =** ◇ + **2**

   A. 1   **B.** 9   **C.** 7   **D.** 10   **E.** 12

3. **?** = **36 ÷ 2**

   A. 18   **B.** 16   **C.** 30   **D.** 38   **E.** 17

4. **21 = 3 x ?**

   A. 19   **B.** 5   **C.** 6   **D.** 7   **E.** 25

5. **? -** ◇ = **33**   ◇ = **17**

   A. 16   **B.** 50   **C.** 20   **D.** 30   **E.** 60

6. **? + 6 = 5 x ◇**   **◇ = 2 x 3**

A. 30   **B.** 24   **C.** 6   **D.** 11   **E.** 20

7. **? = ◇ + 9**   **15 = ◇ - ○**   **○ = 3**

A. 18   **B.** 14   **C.** 27   **D.** 24   **E.** 21

8. **32 ÷ 4 = ? ÷ 5**

A. 8   **B.** 28   **C.** 9   **D.** 35   **E.** 40

9. **? = ◇ - 3**   **○ = ◇ x 2**   **○ = 30**

A. 12   **B.** 15   **C.** 20   **D.** 18   **E.** 16

10. **35 ÷ 5 = 1 + ?**

A. 7   **B.** 30   **C.** 8   **D.** 6   **E.** 7

NUMBER PUZZLES

11. **72 ÷ ? = 6**

   A. 8   **B.** 66   **C.** 12   **D.** 14   **E.** 78

12. **? + 9 = 5 x ◇   8 = 12 - ◇**

   A. 14   **B.** 4   **C.** 2   **D.** 15   **E.** 11

13. **84 = 7 x ?**

   A. 77   **B.** 91   **C.** 14   **D.** 12   **E.** 16

14. **? = 100 ÷ 2**

   A. 50   **B.** 25   **C.** 75   **D.** 10   **E.** 200

15. **? = ◇ - 6   ◯ = ◇ ÷ 6   ◯ = 10**

   A 16   **B.** 54   **C.** 45   **D.** 60   **E.** 66

16. **?** = ◇ **- 21**   **5** = ◇ **÷ 12**

    A. 60   **B.** 50   **C.** 39   **D.** 37   **E.** 40

COGAT Level 11 Grade 5 Answer Explanations

VERBAL BATTERY

Sentence Completion

1. **B.** The word 'microphone' is the best word as this noun is an instrument used to amplify sound and 'project her voice around the room'.

2. **A.** The word 'abolish' is the best word as this verb means to 'formally put an end to something'.

3. **C.** The word 'inhale' is the best word as this verb means to 'breathe in.'

4. **E.** The word 'numerous' is the best word as this adjective means 'great in number'.

5. **A.** The word 'pitchfork' is the best word as this is a tool used for lifting hay.

6. **B.** The word 'extinct' is the best word as this adjective means 'no longer in existence'.

7. **A.** The word 'capsized' is the best word as this verb means to 'overturn in the water.'

8. **E.** The word 'dormitory' is the best word as it refers to a large bedroom for a number of people in a school.

9. **D.** The word 'humble' is the best word as it means to 'be modest of one's own fortune or importance' and means the opposite of 'boastful'.

10. **B.** The word 'afford' is the best word as this verb means 'to have enough money to pay for'.

11. **C.** The word 'dangerous' is the best word as it refers to the fact that snakes can 'cause harm or injury'.

12. **A.** The word 'accurate' is the best word as it refers to an 'exact number', while an estimate refers to an approximate number.

13. **C.** The word 'expel' is the best word as this verb means to 'remove from a school or organization'.

14. **E.** The word 'steeple' is the best word as it refers to the church tower, typically erected from its roof.

15. **B.** The word 'tropics' is the best word as it refers to the natural habitat of parrots.

16. **B.** The word 'upbeat' is the best word as this adjective means 'cheerful' or 'optimistic'.

17. **E.** The word 'perspective' is the best word as this noun means a 'point of view'.

18. **D.** The word 'awaiting' is the best word as this verb refers to waiting for the 'trial' to be held.

19. **D.** The word 'exposed' is the best word as this verb refers to a 'lack of protection against the virus'.

20. **A.** The word 'articulate' is the best word as this verb refers to a person's ability to 'speak fluently and coherently'.

Verbal Classifications

1. **B.** The similarity among the items is they are all marsupials.

2. **A.** The similarity among the items is they are all individual sports rather than team sports.

3. **E.** The similarity among the items is they are all body parts located below the hips.

4, **A.** The similarity among the items is they are all kinds of teeth.

5. **B**. The similarity among the items is they are all parts of the eye.

6. **D**. The similarity among the items is they are all parasites.

7. **B**. The similarity among the items is they are all items of clothing worn that can be worn on the head.

8. **D**. The similarity among the items is they are all people who work with medicine.

9. **D**. The similarity among the items is they all measure time.

10. **A**. The similarity among the items is they are all types of enclosures found on a farm.

11. **C**. The similarity among the items is they all mean 'to make a formal and public declaration.'

12. **B**. The similarity among the items is they are all deserts.

13. **B**. The similarity among the items is they are all mountain ranges.

14. **B**. The similarity among the items is they are all types of jobs.

15. **D**. The similarity among the items is they are all female animals.

16. **B**. The similarity among the items is they are all adjectives that mean 'calm and untroubled.'

17. **A**. The similarity among the items is they are all mythical creatures.

18. **D**. The similarity among the items is they are all types of religions.

19. **C**. The similarity among the items is they are all articles that make up a bed.

20. **A**. The similarity among the items is they are all spices.

Verbal Analogies

1. **B**. A boxer uses gloves as a carpenter uses tools.

2. **C**. Frogs belong to the amphibian class of animals as koalas belong to the marsupial class of animals.

3. **C**. Skiing is performed on a slope as driving is done on a road.

4. **B**. 'Petite' and 'tiny' are synonyms as 'obese' and 'large' are synonyms.

5. **E**. Saturn is an example of a planet as an incisor is a type of tooth.

6. **B**. A turkey is a type of a bird as a cactus is a type of plant.

7. **C**. A cyclist rides a bicycle on a track as a swimmer swims in a pool.

8. **D**. A hairstylist works in a salon as a pilot works in the cockpit of a plane.

9. **D**. A librarian works in a library as a bellhop works in a hotel.

10. **E**. 'Consent' and 'oppose' are antonyms as 'primitive' and 'modern' are antonyms.

11. **C**. 'Bright' and 'brilliant' are synonyms as 'sob' and 'cry' are synonyms.

12. **E**. 'Vacant' and 'empty' are synonyms as 'pupil' and 'student' are synonyms.

13. **D**. A lion can be kept in a cage like valuables can be kept in a vault.

14. **B**. A gas tank holds gas like a bone holds marrow.

15. **A**. 'Provoke' and 'soothe' are antonyms as 'loathe' and 'cherish' are antonyms.

16. **A**. Feet are units of measurement that make up yards and quarts are units of measurement that make up gallons.

17. **D**. 'Debate' and 'agree' are antonyms as 'excavate' and 'bury' are antonyms.

18. **C**. A pencil contains graphite as a thermometer contains mercury.

19. **C**. Rural areas have a lot of open land and soil as urban areas have a lot of roads and pavement.

20. **A**. A breeze is a very light wind and a gale is a very strong wind as a drip is a small drop of liquid and a cascade is a downpour.

21. **B**. A pack is a group of wolves as a pod is a group of whales.

22. **B**. A desk is composed of wood as a tire is composed of rubber.

23. **E**. A joey is a baby kangaroo as a fledgling is a baby bird.

24. **C**. 'Absurd' and 'sensible' are antonyms as 'fatigue' and 'liveliness' are antonyms.

NONVERBAL BATTERY

Figural Classifications

1. **B**. All figures have six sides.

2. **C**. All figures are made up of three shapes.

3. **A**. All figures are four sided shapes with two parts shaded similarly, and two other parts shaded the same.

4. **C**. All shapes have odd numbers of sides.

5. **C**. All shapes have four colors or patterns.

6. **A**. All shapes with even number of sides are shaded black. All shapes with odd number of sides are filled with white.

7. **B**. All figures are only 2-D shapes.

8. **C**. All figures have 5 sides.

9. **C**. All figures have an inner shape shaded black, and an outer shape shaded white.

10. **D**. All figures contain a square that is divided into four equal parts, with a black pentagon pointing upwards and a grey triangle pointing downwards.

11. **B**. All figures contain a square that is divided into nine equal parts, with 4 stars and a rounded rectangle. There is also a diagonal row of 3 stars, a diagonal row that contains 2 stars, and a rounded rectangle at any one of the corners of the bigger square.

12. **A**. All stars are 5 pointed and contain a black rectangle, black star, a grey rectangle, and a grey triangle arranged in four corners of the star. The corner after the triangle (in a clockwise direction from the corner with the triangle) has no shape.

13. **E**. All figures are divided into 4 equal quarters.

14. **B**. All stars are equally divided into halves. The shapes on the left hand side are black and the shapes on the right hand side are grey. The number of sides in the right hand side shape is greater than the number of sides in the left hand side.

15. **D**. All figures have two similar shapes. The bottom shape is black and the top shape is grey. The bottom shape is rotated 90 degrees counterclockwise to form the top shape.

16. **C**. All figures have two shapes. The bottom shape is filled with dots (black or white) and the top shape is filled with horizontal lines.

17. **B**. All figures have two shapes, the outer shape is black and the inner shape is white. The inner shape is touching the outer shape at some points.

18. **E**. All shapes have an outer shape with five sides and an arrow shaded white and grey. The arrow head touches a corner of the shape.

19. **C**. All shapes have a line segment that divides them into two parts. The line segment has arrow heads on both ends. The shapes on both sides of the line segment are the same.

20. **B**. All shapes have 5 different types of arrows arranged in the same order.

21. **A**. All shapes have an outer shape and an inner shape. The number of sides of the inner shape is greater than the number of sides of the outer shape. Answer choice D is incorrect as the outer shape is not a polygon. Answer choice E is incorrect as the inner shape is not a polygon.

22. **D**. All shapes have an outer polygon, a middle ellipse and an inner grey curved object. Answer choice E is incorrect as the inner shape is a circle and the corners of the outer shape are rounded. Answer choice A is incorrect as the outer shape is an ellipse.

Figure Matrices

1. **A**. The shapes are folded inward.

2. **B**. The black arrow folds in first, then the light gray arrow folds in.

3. **C**. The shapes are folded inward.

4. **B**. The correct box completes the mural of two stars in the bottom row while the background shades alternate.

5. **E**. Across each row, the outer circles change shades as the numbers increase by 2, Down the columns, the numbers decrease by 3 from top to bottom.

6. **E**. The black shapes alternate between large, medium and small, while the foreground shapes (medium gray) move counterclockwise around the black shape.

7. **E**. In each row, the squares from column 1 combine with the squares from column 2 to become column 3.

8. **C**. The shape is reduced while the shades are inverted.

9. **E**. The lines complete the spiral mural.

10. **C**. In the second column, the figures are reflected vertically and the colors are inverted.

11. **A**. The figures meet at the horizon and then reflect on the horizon.

12. **B**. The background shapes invert colors/ shades, while the topmost shape remains the same.

13. **A**. The figures are rotated 180 degrees, and the colors/shades are inverted.

14. **D**. Moving from the left to right box in the top row, the shape with three figures decreases by one and the shape with two figures (colored/shaded rings) increases by one. The bottom boxes must reflect the same relationship.

15. **C**. Moving from left to right box, the color of the background and inner shapes are inverted.

16. **E**. In the left box are three triangles and three arrowheads. Moving from the left to right box, the position of the arrowheads are left undisturbed, but the colors/shades of the arrowheads are inverted. The set of three triangles rotates 180 degrees but the triangles' shades/colors are retained.

17. **C**. Moving from the left to right box, the background color/shade changes to black,

the shape at the top moves to the bottom and it is rotated 180 degrees. The shape at the bottom right changes color to white and moves to the top, the shape at the bottom left retains its color/shade and moves to the top.

18. **D**. Moving from the left to right box, the lines form a shape filled with black.

19. **A**. Moving from the left to right box, the black)changes to medium gray,white changes to black, and medium gray changes to white. The size of the left column of shapes remains the same, while the shapes in the middle column becomes medium sized, and the shapes in the right column becomes small sized.

20. **C**. The shapes rotate 80 degrees counterclockwise while the background image increases in size.

21. **D**. The lines in the left column, if surrounded by an outside line, form the shape in the right column.

22. **E**. The bottom right figure completes the larger mural.

Paper Folding

1. **C**. The square paper is folded diagonally in half from the top left corner to bottom right corner. 5 holes are cut on the folded paper. Since the paper is folded once, the holes go through 2 layers of paper. So there should be 10 holes in the answer choice. Answer choices D and E do not have exactly 10 holes. In answer choices A and B, the position of the original holes are different. Hence answer choice C is the correct answer.

2. **E**. The square paper is folded lengthwise in half from the right side to the left side. Four holes are cut on the folded paper. Since the paper is folded once, the holes go through 2 layers of paper. So there should be 8 holes in the answer choice. Since the paper is folded lengthwise the holes on the right side must be the mirror images of the holes of the left. Answer choices B and C do not have exactly 8 holes. In answer choice A, the position of the original holes are different. In answer choice D, the holes on the right are not the mirror images of the holes on the left. Hence answer choice E is the correct answer.

3. **B**. The two corners of the square paper is folded inwards without overlapping. 9 holes are cut on the folded paper, but the holes in the center, lower left corner and upper right corner goes through only one layer of paper. The row of 3 holes go through 2 layers of paper. So there should be (3 + 3 + 1 + 1 + 1) + 6 = 15 holes in the answer choice. Answer choices A, C and E do not have exactly 15 holes. In answer choice D, the position of the original holes are different. Hence answer choice B is the correct answer.

4. **A**. The square paper is folded diagonally in half from the bottom right corner to the top left corner. 5 holes are cut on the folded paper. Since the paper is folded once, the holes go through 2 layers of paper. So there should be 10 holes in the answer choice. Answer choice D does not have exactly 10 holes. In answer choices C and E, the position of the original holes are different. In answer choice B, the holes do not fit the diagonal mirroring pattern. Hence answer choice A is the correct answer.

5. **D**. The square paper is folded inwards towards the middle from the top edge of the paper. Then the bottom left and the right corners are folded towards the center of the square. Then 5 holes are cut on the folded paper. Since the folds do not overlap, the holes go through 2 layers of paper. So there should be 10 holes in the answer choice.

Answer choices C and E do not have exactly 10 holes. In answer choices A and B, the position of the original holes are different. Hence answer choice D is the correct answer.

6. **B.** The square paper is folded diagonally in half from the top left corner to bottom right corner. 5 holes are cut on the folded paper. Since the paper is folded once, the holes go through 2 layers of paper. So there should be 10 holes in the answer choice. Answer choices A, C and E do not have exactly 10 holes. In answer choice D, there are 10 holes but the position of the original holes are different. Hence answer choice B is the correct answer.

7. **E.** The square paper is folded widthwise in half and then lengthwise in half. 3 holes are cut on the folded paper.. Since the paper is folded in half twice, the holes are cut through four layers of paper. So there should be 4 x 3 = 12 holes in the answer choice. Answer choices A, B and C do not have exactly 12 holes. In answer choice D, there are 12 holes, but the arrangement of the holes are not in the correct place given the folding of the paper. There is a slanting row of 3 holes starting from the upper left corner in the folded paper. But there are no holes in the bottom left and right corners. Hence answer choice E is the correct answer.

8. **C.** The square paper is folded inwards towards the middle from the right edge of the paper. Then the top and the bottom corners at the left are folded towards the center of the square. Then 4 holes are cut on the folded paper. The foldings do not overlap. Hence the holes go through 2 layers of paper. So there should be 2 x 4 = 8 holes in the answer choice. Answer choices D and E do not have exactly 8 holes. In answer choices A and B there are 8 holes, but the position of the original holes are different. Hence answer choice C is the correct answer.

9. **A.** The square paper is folded widthwise in half and the top right corner is folded diagonally towards the center of the square. 2 holes are cut on the folded paper. Since the paper is folded twice, the holes go through four layers of paper. So there should be 4 x 2 = 8 holes in the answer choice. Answer choices C, D and E do not have exactly 8 holes. In answer choice B, the position of the original holes are different. Hence answer choice A is the correct answer.

10. **D.** The square paper is folded from each corner towards the center of the square. The folds do not overlap. Hence the holes go through 2 layers of paper. 6 holes are cut on the folded paper. So there should be 2 x 6 = 12 holes in the answer choice. Answer choices A, B and C do not have exactly 12 holes. In answer choice E, the position of the original holes are different. Hence answer choice D is the correct answer.

11. **C.** The square paper is folded diagonally in half from top left corner to bottom right corner. Then the paper is folded from the bottom left corner to the top right corner. Since the paper is folded twice, the holes go through 4 layers of paper. 2 holes are cut on the folded paper.. So there should be 4 x 2 = 8 holes in the answer choice. Answer choices A, D and E do not have exactly 8 holes. In answer choice B, the position of the holes are not in the correct position, given the diagonal folding pattern. Hence answer choice C is the correct answer.

12. **B.** The square paper is folded diagonally in half from bottom left corner to top right corner. Then the paper is folded from the bottom right corner to the top right corner. Since the paper is folded twice, the holes go through 4 layers of paper. 2 holes are cut on the folded paper.. So there should be 4 x 2 = 8 holes in the answer choice. Answer choices

A, C and D do not have exactly 8 holes. In answer choice E, the position of the original holes are different. Hence answer choice B is the correct answer.

13. **E.** The square paper is folded diagonally in half from bottom right corner to top left corner. Then the paper is folded from the upper right corner to the bottom left corner. Since the paper is folded twice, the holes go through 4 layers of paper. 4 holes are cut on the folded paper. So there should be 4 x 4 = 16 holes in the answer choice. Answer choices A, B and D do not have exactly 16 holes. In answer choice C, the position of the original holes are different. Hence answer choice E is the correct answer.

14. **A.** The square paper is folded diagonally in half. A shape is cut on the folded paper. Since the paper is folded once, the shape goes through 2 layers of paper. So there should be 2 shapes in the answer choice. Answer choices B and E have the original shapes in a different position, and answer choice B only has one shape. Answer choice C and D has the second shape in the incorrect place. Hence answer choice A is the correct answer.

15. **B.** The square paper is folded lengthwise in half and then widthwise in half. Three different shapes are cut on the folded paper. Since the paper is folded in half twice, the shapes are cut through four layers of paper. So there should be 4 x 3 = 12 shapes in the answer choice. In answer choice A, C, E, there are 12 shapes, but the arrangement of the holes are not in the correct places given the folding of the paper. Hence answer choice B is the correct answer.

16. **A.** The square paper is folded diagonally in half. Three holes are cut on the folded paper. Since the paper is folded once, the shape goes through 2 layers of paper. So there should be 2 x 3 = 6 shapes in the

correct answer choice. Answer choice D does not have exactly 6 holes. In answer choice B, C, and E, there are 6 holes, but the arrangement of the holes are not in the correct places given the folding of the paper. Hence answer choice A is the correct answer.

QUANTITATIVE BATTERY

Number Analogies

1. **C.** Add 9

2. **D.** Multiply by 6

3. **A.** Divide by 2

4. **E.** Subtract 5

5. **C.** Multiply by 10

6. **D.** Square each number

7. **B.** Divide by 4

8. **A.** Subtract 0 from each number

9. **B.** Square each number

10. **D.** Multiply by 7

11. **B.** Add ¾

12. **E.** Add 5

13. **C.** Subtract ¼

14. **C.** Multiply by 100

15. **A.** Equivalent to ½

16. **A.** Subtract 12

17. **B.** Divide by 3

18. **C.** Multiply by 10

Number Series

1. **E.** Alternatively multiply by 4 and divide by 2. Or, multiply alternating terms (1st, 3rd, 5th etc and 2nd, 4th, 6th etc) by 2.

2. **A.** The 1st, 3rd, 5th terms in the series, etc (odd terms) add 1. The other terms in the series: the 2nd, 4th, 6th, etc (even terms) subtract 1. Or, subtract 3, add 4, subtract 5, add 6, etc.

3. **E.** The 1st, 3rd, 5th terms in the series, etc (odd terms) add 3. The other terms in the series - 2nd, 4th, 6th, etc (even terms) subtract 7.

4. **B.** The 1st, 3rd, 5th terms in the series, etc (odd terms) increase by a number that grows bigger by 1 each time (+3, +4, +5). The other terms in the series -2nd, 4th, 6th, etc (even terms) add 4.

5. **D.** Alternatively add 11 and subtract 3. Or, add 8 to alternating (1st, 3rd, 5th etc and 2nd, 4th, 6th etc) terms.

6. **C.** The 1st, 3rd, 5th terms in the series, etc (odd terms) subtract 0.02. The other terms in the series -2nd, 4th, 6th, etc (even terms) add 0.02.

7. **E.** Alternatively add +2, +2, +2 and -1, -1, -1.

8. **B.** Each term adds 3, 4, 5, 6, etc. progressively.

9. **A.** Alternatively -3, -3, +5.

10. **D.** Add 13 to each term in the series.

11. **C.** The 1st, 3rd, 5th terms in the series (odd terms) add 0.01. The other terms in the series -2nd, 4th, 6th, etc (even terms) add 0.05.

12. **E.** Alternatively subtract 4 and add 12. Or add 8 to alternating (1st, 3rd etc and 2nd, 4th etc) terms.

13. **E.** Alternatively subtract 7 and subtract 9. Or, subtract 16 from alternating (1st, 3rd, 5th etc and 2nd, 4th, 6th etc) terms.

14. **B.** The 1st, 3rd, 5th terms in the series (odd terms) add 5. The other terms in the series - 2nd, 4th, 6th, etc (even terms) add 5.

15. **A.** Progressively add, starting at +3 (+3, +4, +5, + 6, +7, +8, etc.)

16. **C.** Progressively subtract -1, -2, -3, -4, -5, etc.

17. **D.** Alternatively subtract 2 and add 5.

18. **D.** Each set contains three terms. The first term in each set is multiplied by 3 and 4 progressively. (4 x 3 and 4 x 4). The first term in each set adds one progressively. (4, 5, 6, etc.)

Number Puzzles

1. **E.** $25 = 12 + 13$

2. **C.** $7 + 5 = 4 \times 3$ and $5 = 3 + 2$

3. **A.** $18 = 36 \div 2$

4. **D.** $21 = 3 \times 7$

5. **B.** $50 - 17 = 33$

6. **B.** $24 + 6 = 5 \times 6$ and $6 = 2 \times 3$

7. **C.** $27 = 18 + 9$ and $15 = 18 - 3$

8. **E.** $32 \div 4 = 40 \div 5$

9. **A.** $12 = 15 - 3$ and $30 = 15 \times 2$

10. **D.** $35 \div 5 = 1 + 6$

11. **C.** $72 \div 12 = 6$

12. **E.** $11 + 9 = 5 \times 4$ and $8 = 12 - 4$

13. **D.** $84 = 7 \times 12$

14. **A.** $50 = 100 \div 2$

15. **B.** $54 = 60 - 6$ and $10 = 60 \div 6$

16. **C.** $39 = 60 - 21$ and $5 = 60 \div 12$

CogAT® Verbal Battery

Use a No. 2 Pencil
Fill in bubble completely.
Ⓐ ● Ⓒ Ⓓ

Name:_____

Date:_____

1. Ⓐ Ⓑ Ⓒ Ⓓ Ⓔ	1. Ⓐ Ⓑ Ⓒ Ⓓ Ⓔ	1. Ⓐ Ⓑ Ⓒ Ⓓ Ⓔ
2. Ⓐ Ⓑ Ⓒ Ⓓ Ⓔ	2. Ⓐ Ⓑ Ⓒ Ⓓ Ⓔ	2. Ⓐ Ⓑ Ⓒ Ⓓ Ⓔ
3. Ⓐ Ⓑ Ⓒ Ⓓ Ⓔ	3. Ⓐ Ⓑ Ⓒ Ⓓ Ⓔ	3. Ⓐ Ⓑ Ⓒ Ⓓ Ⓔ
4. Ⓐ Ⓑ Ⓒ Ⓓ Ⓔ	4. Ⓐ Ⓑ Ⓒ Ⓓ Ⓔ	4. Ⓐ Ⓑ Ⓒ Ⓓ Ⓔ
5. Ⓐ Ⓑ Ⓒ Ⓓ Ⓔ	5. Ⓐ Ⓑ Ⓒ Ⓓ Ⓔ	5. Ⓐ Ⓑ Ⓒ Ⓓ Ⓔ
6. Ⓐ Ⓑ Ⓒ Ⓓ Ⓔ	6. Ⓐ Ⓑ Ⓒ Ⓓ Ⓔ	6. Ⓐ Ⓑ Ⓒ Ⓓ Ⓔ
7. Ⓐ Ⓑ Ⓒ Ⓓ Ⓔ	7. Ⓐ Ⓑ Ⓒ Ⓓ Ⓔ	7. Ⓐ Ⓑ Ⓒ Ⓓ Ⓔ
8. Ⓐ Ⓑ Ⓒ Ⓓ Ⓔ	8. Ⓐ Ⓑ Ⓒ Ⓓ Ⓔ	8. Ⓐ Ⓑ Ⓒ Ⓓ Ⓔ
9. Ⓐ Ⓑ Ⓒ Ⓓ Ⓔ	9. Ⓐ Ⓑ Ⓒ Ⓓ Ⓔ	9. Ⓐ Ⓑ Ⓒ Ⓓ Ⓔ
10. Ⓐ Ⓑ Ⓒ Ⓓ Ⓔ	10. Ⓐ Ⓑ Ⓒ Ⓓ Ⓔ	10. Ⓐ Ⓑ Ⓒ Ⓓ Ⓔ
11. Ⓐ Ⓑ Ⓒ Ⓓ Ⓔ	11. Ⓐ Ⓑ Ⓒ Ⓓ Ⓔ	11. Ⓐ Ⓑ Ⓒ Ⓓ Ⓔ
12. Ⓐ Ⓑ Ⓒ Ⓓ Ⓔ	12. Ⓐ Ⓑ Ⓒ Ⓓ Ⓔ	12. Ⓐ Ⓑ Ⓒ Ⓓ Ⓔ
13. Ⓐ Ⓑ Ⓒ Ⓓ Ⓔ	13. Ⓐ Ⓑ Ⓒ Ⓓ Ⓔ	13. Ⓐ Ⓑ Ⓒ Ⓓ Ⓔ
14. Ⓐ Ⓑ Ⓒ Ⓓ Ⓔ	14. Ⓐ Ⓑ Ⓒ Ⓓ Ⓔ	14. Ⓐ Ⓑ Ⓒ Ⓓ Ⓔ
15. Ⓐ Ⓑ Ⓒ Ⓓ Ⓔ	15. Ⓐ Ⓑ Ⓒ Ⓓ Ⓔ	15. Ⓐ Ⓑ Ⓒ Ⓓ Ⓔ
16. Ⓐ Ⓑ Ⓒ Ⓓ Ⓔ	16. Ⓐ Ⓑ Ⓒ Ⓓ Ⓔ	16. Ⓐ Ⓑ Ⓒ Ⓓ Ⓔ
17. Ⓐ Ⓑ Ⓒ Ⓓ Ⓔ	17. Ⓐ Ⓑ Ⓒ Ⓓ Ⓔ	17. Ⓐ Ⓑ Ⓒ Ⓓ Ⓔ
18. Ⓐ Ⓑ Ⓒ Ⓓ Ⓔ	18. Ⓐ Ⓑ Ⓒ Ⓓ Ⓔ	18. Ⓐ Ⓑ Ⓒ Ⓓ Ⓔ
19. Ⓐ Ⓑ Ⓒ Ⓓ Ⓔ	19. Ⓐ Ⓑ Ⓒ Ⓓ Ⓔ	19. Ⓐ Ⓑ Ⓒ Ⓓ Ⓔ
20. Ⓐ Ⓑ Ⓒ Ⓓ Ⓔ	20. Ⓐ Ⓑ Ⓒ Ⓓ Ⓔ	20. Ⓐ Ⓑ Ⓒ Ⓓ Ⓔ
		21. Ⓐ Ⓑ Ⓒ Ⓓ Ⓔ
		22. Ⓐ Ⓑ Ⓒ Ⓓ Ⓔ
		23. Ⓐ Ⓑ Ⓒ Ⓓ Ⓔ
		24. Ⓐ Ⓑ Ⓒ Ⓓ Ⓔ

CogAT® Nonverbal Battery

Use a No. 2 Pencil
Fill in bubble completely.

Ⓐ ● Ⓒ Ⓓ

Name:_____

Date:_____

1. Ⓐ Ⓑ Ⓒ Ⓓ Ⓔ	1. Ⓐ Ⓑ Ⓒ Ⓓ Ⓔ	1. Ⓐ Ⓑ Ⓒ Ⓓ Ⓔ
2. Ⓐ Ⓑ Ⓒ Ⓓ Ⓔ	2. Ⓐ Ⓑ Ⓒ Ⓓ Ⓔ	2. Ⓐ Ⓑ Ⓒ Ⓓ Ⓔ
3. Ⓐ Ⓑ Ⓒ Ⓓ Ⓔ	3. Ⓐ Ⓑ Ⓒ Ⓓ Ⓔ	3. Ⓐ Ⓑ Ⓒ Ⓓ Ⓔ
4. Ⓐ Ⓑ Ⓒ Ⓓ Ⓔ	4. Ⓐ Ⓑ Ⓒ Ⓓ Ⓔ	4. Ⓐ Ⓑ Ⓒ Ⓓ Ⓔ
5. Ⓐ Ⓑ Ⓒ Ⓓ Ⓔ	5. Ⓐ Ⓑ Ⓒ Ⓓ Ⓔ	5. Ⓐ Ⓑ Ⓒ Ⓓ Ⓔ
6. Ⓐ Ⓑ Ⓒ Ⓓ Ⓔ	6. Ⓐ Ⓑ Ⓒ Ⓓ Ⓔ	6. Ⓐ Ⓑ Ⓒ Ⓓ Ⓔ
7. Ⓐ Ⓑ Ⓒ Ⓓ Ⓔ	7. Ⓐ Ⓑ Ⓒ Ⓓ Ⓔ	7. Ⓐ Ⓑ Ⓒ Ⓓ Ⓔ
8. Ⓐ Ⓑ Ⓒ Ⓓ Ⓔ	8. Ⓐ Ⓑ Ⓒ Ⓓ Ⓔ	8. Ⓐ Ⓑ Ⓒ Ⓓ Ⓔ
9. Ⓐ Ⓑ Ⓒ Ⓓ Ⓔ	9. Ⓐ Ⓑ Ⓒ Ⓓ Ⓔ	9. Ⓐ Ⓑ Ⓒ Ⓓ Ⓔ
10. Ⓐ Ⓑ Ⓒ Ⓓ Ⓔ	10. Ⓐ Ⓑ Ⓒ Ⓓ Ⓔ	10. Ⓐ Ⓑ Ⓒ Ⓓ Ⓔ
11. Ⓐ Ⓑ Ⓒ Ⓓ Ⓔ	11. Ⓐ Ⓑ Ⓒ Ⓓ Ⓔ	11. Ⓐ Ⓑ Ⓒ Ⓓ Ⓔ
12. Ⓐ Ⓑ Ⓒ Ⓓ Ⓔ	12. Ⓐ Ⓑ Ⓒ Ⓓ Ⓔ	12. Ⓐ Ⓑ Ⓒ Ⓓ Ⓔ
13. Ⓐ Ⓑ Ⓒ Ⓓ Ⓔ	13. Ⓐ Ⓑ Ⓒ Ⓓ Ⓔ	13. Ⓐ Ⓑ Ⓒ Ⓓ Ⓔ
14. Ⓐ Ⓑ Ⓒ Ⓓ Ⓔ	14. Ⓐ Ⓑ Ⓒ Ⓓ Ⓔ	14. Ⓐ Ⓑ Ⓒ Ⓓ Ⓔ
15. Ⓐ Ⓑ Ⓒ Ⓓ Ⓔ	15. Ⓐ Ⓑ Ⓒ Ⓓ Ⓔ	15. Ⓐ Ⓑ Ⓒ Ⓓ Ⓔ
16. Ⓐ Ⓑ Ⓒ Ⓓ Ⓔ	16. Ⓐ Ⓑ Ⓒ Ⓓ Ⓔ	16. Ⓐ Ⓑ Ⓒ Ⓓ Ⓔ
17. Ⓐ Ⓑ Ⓒ Ⓓ Ⓔ	17. Ⓐ Ⓑ Ⓒ Ⓓ Ⓔ	
18. Ⓐ Ⓑ Ⓒ Ⓓ Ⓔ	18. Ⓐ Ⓑ Ⓒ Ⓓ Ⓔ	
19. Ⓐ Ⓑ Ⓒ Ⓓ Ⓔ	19. Ⓐ Ⓑ Ⓒ Ⓓ Ⓔ	
20. Ⓐ Ⓑ Ⓒ Ⓓ Ⓔ	20. Ⓐ Ⓑ Ⓒ Ⓓ Ⓔ	
21. Ⓐ Ⓑ Ⓒ Ⓓ Ⓔ	21. Ⓐ Ⓑ Ⓒ Ⓓ Ⓔ	
22. Ⓐ Ⓑ Ⓒ Ⓓ Ⓔ	22. Ⓐ Ⓑ Ⓒ Ⓓ Ⓔ	

CogAT® Quantitative Battery

Use a No. 2 Pencil
Fill in bubble completely.
Ⓐ ● Ⓒ Ⓓ

Name:_____

Date:_____

1. Ⓐ Ⓑ Ⓒ Ⓓ Ⓔ	1. Ⓐ Ⓑ Ⓒ Ⓓ Ⓔ	1. Ⓐ Ⓑ Ⓒ Ⓓ Ⓔ
2. Ⓐ Ⓑ Ⓒ Ⓓ Ⓔ	2. Ⓐ Ⓑ Ⓒ Ⓓ Ⓔ	2. Ⓐ Ⓑ Ⓒ Ⓓ Ⓔ
3. Ⓐ Ⓑ Ⓒ Ⓓ Ⓔ	3. Ⓐ Ⓑ Ⓒ Ⓓ Ⓔ	3. Ⓐ Ⓑ Ⓒ Ⓓ Ⓔ
4. Ⓐ Ⓑ Ⓒ Ⓓ Ⓔ	4. Ⓐ Ⓑ Ⓒ Ⓓ Ⓔ	4. Ⓐ Ⓑ Ⓒ Ⓓ Ⓔ
5. Ⓐ Ⓑ Ⓒ Ⓓ Ⓔ	5. Ⓐ Ⓑ Ⓒ Ⓓ Ⓔ	5. Ⓐ Ⓑ Ⓒ Ⓓ Ⓔ
6. Ⓐ Ⓑ Ⓒ Ⓓ Ⓔ	6. Ⓐ Ⓑ Ⓒ Ⓓ Ⓔ	6. Ⓐ Ⓑ Ⓒ Ⓓ Ⓔ
7. Ⓐ Ⓑ Ⓒ Ⓓ Ⓔ	7. Ⓐ Ⓑ Ⓒ Ⓓ Ⓔ	7. Ⓐ Ⓑ Ⓒ Ⓓ Ⓔ
8. Ⓐ Ⓑ Ⓒ Ⓓ Ⓔ	8. Ⓐ Ⓑ Ⓒ Ⓓ Ⓔ	8. Ⓐ Ⓑ Ⓒ Ⓓ Ⓔ
9. Ⓐ Ⓑ Ⓒ Ⓓ Ⓔ	9. Ⓐ Ⓑ Ⓒ Ⓓ Ⓔ	9. Ⓐ Ⓑ Ⓒ Ⓓ Ⓔ
10. Ⓐ Ⓑ Ⓒ Ⓓ Ⓔ	10. Ⓐ Ⓑ Ⓒ Ⓓ Ⓔ	10. Ⓐ Ⓑ Ⓒ Ⓓ Ⓔ
11. Ⓐ Ⓑ Ⓒ Ⓓ Ⓔ	11. Ⓐ Ⓑ Ⓒ Ⓓ Ⓔ	11. Ⓐ Ⓑ Ⓒ Ⓓ Ⓔ
12. Ⓐ Ⓑ Ⓒ Ⓓ Ⓔ	12. Ⓐ Ⓑ Ⓒ Ⓓ Ⓔ	12. Ⓐ Ⓑ Ⓒ Ⓓ Ⓔ
13. Ⓐ Ⓑ Ⓒ Ⓓ Ⓔ	13. Ⓐ Ⓑ Ⓒ Ⓓ Ⓔ	13. Ⓐ Ⓑ Ⓒ Ⓓ Ⓔ
14. Ⓐ Ⓑ Ⓒ Ⓓ Ⓔ	14. Ⓐ Ⓑ Ⓒ Ⓓ Ⓔ	14. Ⓐ Ⓑ Ⓒ Ⓓ Ⓔ
15. Ⓐ Ⓑ Ⓒ Ⓓ Ⓔ	15. Ⓐ Ⓑ Ⓒ Ⓓ Ⓔ	15. Ⓐ Ⓑ Ⓒ Ⓓ Ⓔ
16. Ⓐ Ⓑ Ⓒ Ⓓ Ⓔ	16. Ⓐ Ⓑ Ⓒ Ⓓ Ⓔ	16. Ⓐ Ⓑ Ⓒ Ⓓ Ⓔ
17. Ⓐ Ⓑ Ⓒ Ⓓ Ⓔ	17. Ⓐ Ⓑ Ⓒ Ⓓ Ⓔ	
18. Ⓐ Ⓑ Ⓒ Ⓓ Ⓔ	18. Ⓐ Ⓑ Ⓒ Ⓓ Ⓔ	

BONUS
QUANTITATIVE CHALLENGE QUESTIONS

If you also want additional quantitative challenge questions, please go to the following link to download them!

To get your challenge questions today, please visit:
https://originstutoring.lpages.co/cogat-11-challenge-questions/

Challenge questions can help a student get used to doing the most difficult questions on the test.

Made in the USA
Coppell, TX
28 May 2024

32875814R00055